Power Distribution Engineering

Additional Volumes in Preparation

Modern Digital Control Systems, Second Edition, *Raymond G. Jacquot*

Integrated Circuit Quality and Reliability, Second Edition, Revised and Expanded, *Eugene R. Hnatek*

Handbook of Electric Motors, *edited by Richard Engelmann and William H. Middendorf*

Adaptive IIR Filtering in Signal Processing and Control, *Philip Regalia*

Power Distribution Engineering

Fundamentals and Applications

James J. Burke
Power Technologies, Inc.
Schenectady, New York

MARCEL DEKKER, INC.   NEW YORK · BASEL

Library of Congress Cataloging-in-Publication Data

Burke, James J.
    Power distribution engineering: fundamentals and applications/James J. Burke.
        p.   cm.—(Electrical engineering and electronics)
    Includes bibliographical references and index.
    ISBN 0-8247-9237-8 (alk. paper)
    1. Electrical power distribution.   I. Title.   II. Series.
TK3001.B87 1994
621.319—dc20   94-11103
                                                                              CIP

The publisher offers discounts on this book when ordered in bulk quantities. For more information, write to Special Sales/Professional Marketing at the address below.

This book is printed on acid-free paper.

MARCEL DEKKER, INC.
270 Madison Avenue, New York, New York 10016

Current printing (last digit):

20 19 18 17 16 15 14 13 12

PRINTED IN THE UNITED STATES OF AMERICA

To the memory of Aunt Marg & Uncle Rick

PREFACE

The purpose of this book is to give the utility distribution engineer a document which is truly useful and up-to-date for today's competitive environment. The book is based on over 25 years of experience in both performing distribution studies and in teaching courses in distribution engineering. Too often the author has found that books on distribution engineering are either simply very good reference data books or good academic textbooks. Neither document usually explains why utilities do things the way they do. It is hoped that this book will bridge that gap.

Simplicity of concepts is emphasized. Complex mathematical concepts are not used since teaching experience has shown that understanding concepts is far more valuable than being able to use a wide variety of mathematical techniques. The book's math is straightforward and is used only to illustrate the point that many seemingly difficult concepts and calculations can be greatly simplified using obvious assumptions.

This book is intended to cover a wide and comprehensive list of topics, from system protection to economic evaluations. Its chapters reflect current and future thinking in these areas and are based on the author's work in IEEE Standards groups as well as many real utility studies.

The material varies in depth. For example, Chapter 1 is meant to be a brief overview of a distribution system. This was done because there are several good documents presently on the market that address the topic in greater detail. On the other hand, Chapters 4 and 5, on overcurrent and overvoltage protection, are quite comprehensive. This was done for two reasons. First, philosophies in both these areas are in a state of flux. Second, very little material on either of these areas is currently available in a comprehensive form.

Chapter 7, on power quality, appears near the end of the book since the concept of power quality can mean just about anything and as such will be more meaningful if the earlier chapters are read beforehand. Typical topics, such as sags, swells, and harmonics, are covered as are other areas such as electromagnetic fields and stray voltage, considered by only some distribution engineers to be associated with power quality.

Each chapter has review questions meant to stimulate thought for those readers who utilize the book as a self-teaching guide. Although this book can serve as a good day-to-day reference, gaining the maximum value from its contents will be helped by attempting these exercises. The questions are meant to be simple and fun.

The author would like to acknowledge help over the years from Harold Campbell as well as the late Jack Easley, Norm Schultz, and Bill Moody. Finally, the author wishes to thank Sherie Roseboom, who has worked so hard and so long typing and assembling the many drafts and manuscripts.

JAMES J. BURKE

CONTENTS

Power
Distribution
Engineering

1

UTILITY DISTRIBUTION SYSTEM DESIGN AND CHARACTERISTICS

INTRODUCTION

The distribution engineer sometimes finds it difficult to define a typical distribution system. It is the purpose of this chapter to suggest typical values of voltage, line lengths, load and fault levels, as well as types of system design and grounding which can be used as background information for the more technical discussions later on in the book.

DESIGN

The Utility System

The electric utility system is usually divided into three segments which are generation, transmission, and distribution. A fourth division, which can sometimes be made is subtransmission, which can really be considered a subset of transmission since the voltage levels overlap and operational and protection practices are quite similar. Figure 1-1, shown below, illustrates some of the major components in these divisions.

The distribution system, which is our main area of interest, is commonly broken down into the following three components:

1. Distribution substation
2. Distribution primary
3. Secondary

Even on this greatly simplified one-line diagram, it can be seen that the distribution system consists of a much wider variety of voltage levels, components, loads and interconnections than does the generation or transmission system.

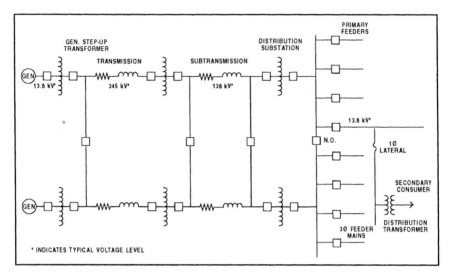

Figure 1-1. Typical Electric Supply System

Distribution Substation

The distribution system is fed through distribution substations. These substations consist of an almost infinite number of designs based on considerations such as

- Load density
- High side voltage
- Low side voltage
- Land availability
- Reliability requirements
- Load growth
- Voltage drop
- Emergency conditions
- Cost and losses

A typical substation is shown in Figure 1-2. This substation indicates the average arrangement and equipment ratings per an industry survey. For example, the voltage of the high side bus can be anywhere from 34.5 kV all the way up to 345 kV and beyond. The average or preferred high side voltage level is approximately 115 to 138 kV because this voltage level is usually high enough to maintain a "stiff" enough source and low enough to alleviate the costs associated with high side equipments. As shown, the average substation consists of two transformers rated 21/28/35 MVA (OA/FA/FOA) with an impedance of approximately 10 percent. Protection of the substation transformer is usually attained via high side breakers or circuit switchers and low side breakers used in conjunction with differential relays (overcurrent relaying is also used as backup and is not shown). The transformer low side breaker, sometimes referred to as the "substation secondary breaker", is also used to protect against low voltage bus faults as well as back up the feeder breakers.

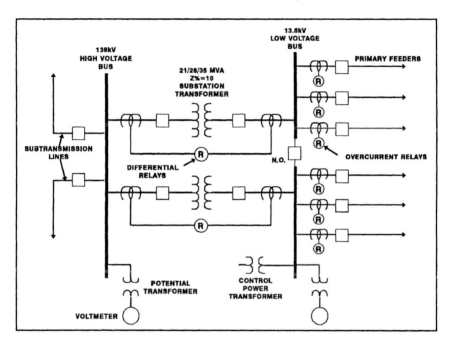

Figure 1-2. Average Distribution Substation Arrangement

The low voltage bus in a multiple transformer substation is usually split (contains a normally open breaker or switch) to alleviate circulating currents as well as reduce the short circuit current seen by the system. Two or more feeders are normally connected to each bus through a feeder breaker. On smaller substations where short circuit levels are lower, a recloser is sometimes used instead of a breaker. Short circuit levels at the

3

terminals of the low voltage bus are generally kept at 12,000 amperes or less although there are many systems where much higher levels can be found.

Distribution Feeders

Figure 1-3 shows a primary distribution feeder with various equipment such as fuses, distribution transformers, reclosers, and switches connected to it. Much of this equipment, such as a recloser, is utilized only at the distribution level. On the other hand, some of the equipment such as capacitors, transformers, and arresters is also used at the transmission levels but with considerably different rules of application. As shown, most distribution feeders are 3-phase and 4-wire. The fourth wire is the neutral wire which is connected to the pole, usually below the phase wires, and grounded periodically.

A three-phase feeder main can be fairly short, on the order of a mile or two, or it can be as long as 30 miles. Voltage levels can be as high as 34.5 kV, with the most common voltages being in the 15 kV class. While most of the 3-phase mains are overhead, much of the new construction, particularly the single-phase lateral construction, is being put underground. Underground systems have the advantage of immunity from certain types of temporary fault conditions like wind, direct lightning strikes, animals, etc. Permanent faults, on the other hand, are much more difficult to locate and repair and have been the subject of much concern in recent years.

There are various feeder designs which can increase customer service. Five of these are shown below in Figure 1-4. These designs are described as follows:

Radial System. It is obvious that the radial system is exposed to many interruption possibilities, the most important of which are those due to primary overhead or underground cable failure or transformer failure. Either event may be accompanied by a long interruption, given nominally by some utilities as 10 to 12 hours. Both components have finite failure rates and such interruptions are expected and statistically predictable. The system will be satisfactory <u>only</u> if the interruptions frequency is very low and if there are ways to operate the system without planned outages. Feeder breakers reclosing or temporary faults may affect sensitive loads.

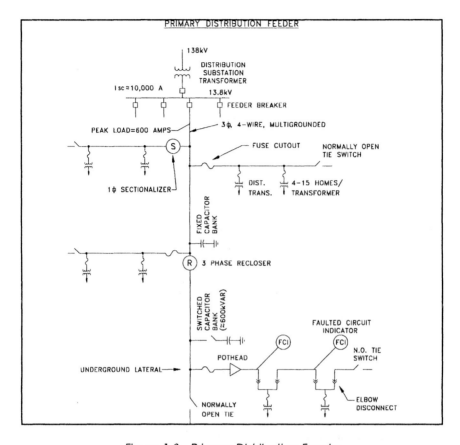

Figure 1-3. Primary Distribution Feeder

Primary Loop. A great improvement is obtained by arranging a primary loop, which provides two-way feed at each transformer. In this manner, any section of the primary can be isolated, without interruption, and primary faults are reduced in duration to the time required to locate a fault and do the necessary switching to restore service. This procedure can be performed either manually or automatically. The cable in each half of the loop must have capacity enough to carry all the load. The additional cable exposure will tend to increase the frequency of faults, but not necessarily the faults per customer. The addition of a loop tie switch at the open point also introduces the possibility of a single equipment fault causing an interruption to both halves of the loop. Murphy's Law

5

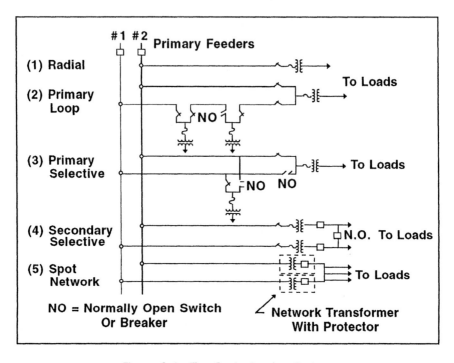

Figure 1-4. Five Basic Service Systems

generally applies to these situations. Once again, sensitive loads would be affected by reclosing under temporary fault conditions.

Primary Selective. This system uses the same basic components as in the primary loop, but arranged in a dual or main/alternative scheme. Each transformer can "select" its source, and automatic switching is frequently used. When automatic, the interruption duration can be limited to two or three seconds. Each service now represents a potential two-feeder outage (if the open switch fails), but under normal contingencies, service restoration is rapid and there is no need to locate the fault (as with the loop) prior to doing the switching. This scheme is in popular use on many underground systems. It also offers little remedy for computer problems caused by temporary faults to the overhead system.

Secondary Selective. This is the first of the service systems using two transformers and low voltage switching. It is not in popular use by utilities for 480 volt service, but is common in industrial plants and on institutional properties. Primary operational switching is eliminated and with it some causes of difficulty. Duplicate transformers virtually eliminate the possibility of a long interruption due to failure. Load is divided between the two units and automatic transfer is employed on loss of voltage to either load. There must be close coordination of utility and customer during planned transfers, and the split responsibility is probably the principal reason for its limited use as a service system. Temporary faults on the primary feeders should have little if any effect on even sensitive computer loads.

Secondary Spot Network. Maximum service reliability and operating flexibility are obtained by use of the spot network using two or more transformer/protector units in parallel. The low voltage bus is continuously energized by all units, and automatic disconnection of any unit is obtained by sensitive reverse power relays in the protector. Maintenance switching of primary feeders can be done without customer interruption or involvement. Spot networks are common in downtown, high density areas and are being applied frequently in outlying areas for large commercial services where the supply feeders can be made available. This system also represents the most compact and reliable arrangement of components and is the most reliable for all classes of loads.

Each of the five systems described can be evaluated in terms of reliability for traditional loads as shown below. As can be seen (see Table 1-1), a radial system, which is used in some residential areas, can expect an outage almost once a year lasting approximately 90 minutes. On the other hand, a spot network, which many utilities use in downtown areas, will see only two outages for every 100 years.

Network Systems

Secondary ac network systems (grid networks) began around the year 1915 replacing the older dc networks which had problems with cost of converters, copper costs and voltage difficulties. Most cities (over 260) in the U.S. have an ac network system. These network systems are well known for their high reliability (see Table 1-1), although their cost differential has favored the primary selective design for new construction in recent years.

Table 1-1. Measured Reliability of Different Distribution Systems							
Type of System	Radial	Primary Auto-Loop	URD	Primary Selective	Secondary Selective	Grid Network	Spot Network
Outages/yr	0.3-1.3	0.4-0.7	0.4-0.7	0.1-0.5	0.1-0.5	0.005-0.020	0.02-0.10
Average outage duration, min	90	65	60	180	180	135	180
Momentary interruptions/yr	5-10	10-15	4-8	4-8	2-4	0	0-1

The major segments of a network system, as shown in Figure 1-5, are:

- Primary feeder circuits
- Network units (consisting of the network transformer and the network protector)
- Secondary grid

The secondary grid is either 208Y/120 or 480Y/277 with virtually all spot networks favoring the higher voltage (see Figure 1-6). Commonly used wire sizes range from 4/0 to 500 MCM AWG. The main protection of the secondary grid comes from the ability of the system to "burn off" the fault. This, of course, meant that no protective device was required to operate. While this practice proved to be highly successful on the 208Y/120 volt secondary network, there have been many instances of the 480Y/277 volt system not being able to successfully burn itself clear, sometimes resulting in fires and considerable damage. A partial solution to this has been the use of "limiters", which are devices (really just restricted copper sections which act like a fuse) installed in the secondary main at each junction point. It should be noted that the "limiter" is usually like an expulsion fuse and does not limit the magnitude of current like a true current limiting fuse. In high short circuit areas, true "current limiting" fuses are now being used to limit damage due to these secondary faults.

The network protector is an electrically operated low-voltage air circuit breaker with self-contained relays for controlling its operation. Its main purpose is to isolate the secondary from problems on the source side

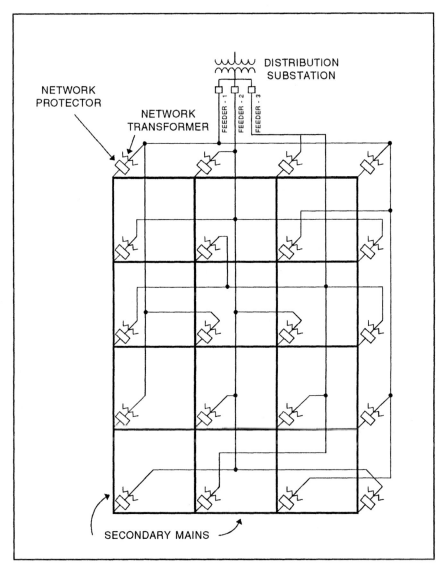

Figure 1-5. Single-Phase, Three-Wire Secondary
System 240/120 Volt Service

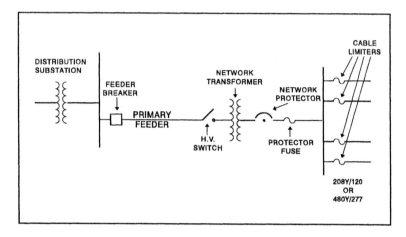

Figure 1-6.

(network transformer, primary feeder, etc.). It is not its function to isolate faults on the secondary network itself. The network protector performs the following three basic functions:

1. Provides automatic isolation of faults in the primary feeders or network transformers.
2. Provides ability to trip protector on reverse power.
3. Provides ability to close automatically when feeder is energized, and in-phase or leading.

The network protector is equipped with a set of fuses, one in each phase, located between the circuit breaker and the terminals for connection to the system secondary voltage grid. The primary function of the fuse is to provide backup to the protector, not to clear secondary faults. These secondary faults are either supposed to "burn clear" or be cleared by the limiters.

Secondaries

The purpose of the distribution transformer is to reduce the primary voltage to a level where it can be utilized by the customer. Three-phase commercial distribution transformers range in size anywhere from about 75 kVA to over 2000 kVA. Single-phase transformers range in size from about 10 kVA to about 300 kVA with units in the 25 and 37.5 kVA size being the most popular for residential areas.

The secondary voltage level in the United States for residential service is 120/240. What this means is that a typical residence has a choice of either depending on the requirements of the load. Figure 1-7 illustrates the connection of a typical single-phase, three-wire secondary system used in most homes. Typically, lower wattage devices (lights for example) are connected line-to-neutral across both sides of the transformer secondary. Higher wattage devices such as ovens, clothes dryers, etc., are usually connected across the 240 volt circuit since this has the effect of reducing voltage drop and losses.

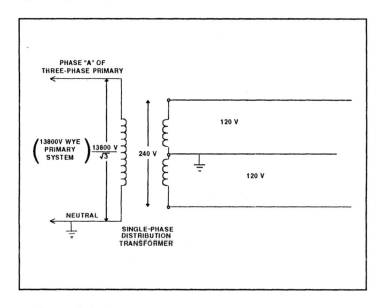

*Figure 1-7. Single-Phase, Three-Wire Secondary System
240/120 Volt Service*

Figure 1-8 shows the connections for a three-phase, four-wire secondary system rated 208/120 volts. This type of connection is used where moderate three-phase loading as well as single-phase load is prevalent. For light industrial and commercial loads with higher power requirements, 480/277 is utilized. Secondary network systems use either 208/120 or 480/277 depending on load requirements.

11

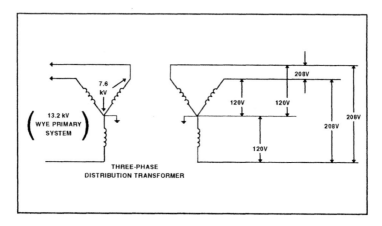

Figure 1-8. Three-Phase, Four-Wire Secondary System 208/120 Volt Service

OVERHEAD VS. UNDERGROUND

Reliability

Underground distribution has become commonplace. The major drive from overhead distribution to underground distribution was primarily a response to environmental pressures. Most utilities today put most of their single-phase residential developments (5 homes or more) underground. Many utilities are also very gradually converting existing three-phase overhead to underground.

Most engineers agree that underground is more reliable than overhead. The argument is many times made that underground systems fail less often but take much longer to fix. Also, the older URD cable has been failing at considerably higher rates than expected. The bottom line is that "average customer minutes outage" (CMO), is considerably lower for most URD designs. Undergrounding also eliminates almost all temporary faults which, for some systems, is 80% of their total. In these days of "power quality", where momentary outages are a large concern, URD can have a very desirable effect. A comparison of the failure rates (permanent faults) between overhead and underground is shown below.

Table 1-2		
	Failure Rate (Failure/Year/Mile)	
Voltage Level	Overhead Line	Underground Cable
5 to 11 kV	0.177	0.048
11 to 20 kV	0.130	0.097
33 kV	0.070	0.037
66 kV	0.059	0.028

Equipment

Most equipment used in an overhead distribution system utilizes air as its primary insulating medium (approximate 186 kV per foot for an impulse). For example, most overhead lines are uninsulated or in the case of "tree wire", minimally insulated. Most overhead switches such as the standard disconnects rely on air to insulate the open contacts or even break very low level currents such as low charging currents. Air-break switches use air to break load currents but these switches are essentially disconnect switches equipped with arcing horns.

Underground equipment is different because the ground is in such close proximity and air insulation is now usually insufficient. Phase conductors must consequently be insulated (cable) and certain other equipments such as switches must utilize a more effective insulating medium such as vacuum, oil and most recently, SF6. A new term associated with underground distribution is "deadfront". The term "deadfront" (e.g., deadfront switch) means that there are no exposed connections. For example, when the cabinet door of a deadfront switch is opened, no "live parts will be exposed". The loadbreak elbow when connected into the bushing, shown in Figure 1-9, would be classified a "deadfront system". On the other hand, most overhead equipment and some URD equipment is classified as "livefront" because exposed parts like spade connectors are clearly visible.

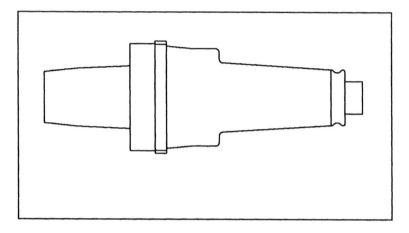

Figure 1-9. Integrated Bushing

Some terminology common to overhead and underground switches which tends to cause confusion is defined unofficially (i.e., in layman's terms, not IEEE) as follows:

a.   Loadbreak. Loadbreak means the device has the ability to break load; usually 200 amps maximum for single-phase and 600 amperes maximum for 3∅ phase. Many overhead switches have no loadbreak capability while most underground switches are, in part, rated for breaking load.

The term "loadbreak" in URD is normally associated with the "loadbreak elbow" shown in Figure 1-10. The loadbreak elbow is actually a mini, single-phase switch that allows the system to be sectionalized (opened) under energized conditions where loads up to 200 amperes on the laterals can be interrupted. A few utilities still use what is called a "non-loadbreak elbow". This elbow looks similar but cannot be switched under load (energized) conditions.

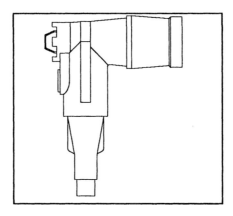

Figure 1-10. Molded-shield Elbows

b. Continuous Rating. The maximum ampere rating of the device under continuous operation. If the device is a switch, "continuous" does not mean that the switch can interrupt this load; it just means it can pass this load, in the closed position, without damage.

c. Momentary Rating. The "momentary rating" of a device is the amount of short circuit current it can pass, in the closed position, without damage (still be operable). It does not mean the device can interrupt the fault current. For example, a loadbreak elbow has a momentary rating and cannot be used to interrupt anything higher than load current (generally 200 amperes).

d. Short Circuit Rating. The short circuit rating of a device is the maximum current it is designed to interrupt. Examples of devices with this rating are fuses, breakers and reclosers.

e. Close and Latch. The "close and latch" rating of a switch is the maximum ampere rating (fault level) that the switch can close into successfully. It is never normal practice to close into a fault. However, by mistake, it is entirely possible that a switch will be closed into a fault. The "close and latch" capability of a device is meant to protect the operator from this error. Even simple switches, like the loadbreak elbow, have close and latch ratings. If an elbow is closed into a fault, within its rating, it will survive but it should be replaced.

f. BIL. BIL (basic impulse level) is a rating which allows the user to assess the voltage impulse capability (ability to withstand impulses

15

without failure) of the equipment. To establish this rating, the equipment is tested with a voltage impulse wave defined as 1.2 × 50 μs. An example will help illustrate what all this means. Equipment on a 13.8 kV system normally has a BIL of 95 kV. This means that this equipment has been impulsed with a wave (see Figure 1-11) which rises to a crest value of 95 kV in 1.2 microseconds and decays to half its value in 50 microseconds.

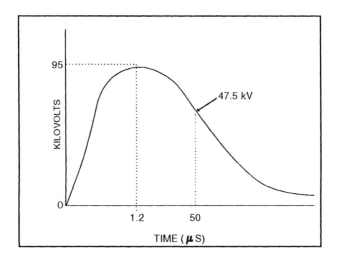

Figure 1-11. Standard BIL Wave

GROUNDING

Grounding on the primary distribution system (2400 volts to 34.5 kV) is usually either 4-wire multigrounded or delta. A 4-wire multigrounded system, which is by far the most popular, means that the substation is grounded and a fourth wire, the neutral wire, is carried along with the phase wire and grounded periodically (see Figure 1-12(a)). Some utilities ground as little as four times per mile while others ground at every pole. Sometimes the substation's transformer is grounded through an impedance (approximately 1 ohm) in an effort to limit short circuit current levels. Some of the more important advantages of a 4-wire multigrounded system (over a 3-wire delta) are:

1. High short circuit currents allowing effective overcurrent relaying practice.
2. Much cheaper for single-phase service, especially underground, since only one cable, bushing, switch, fuse, etc., needs to be used as compared to a delta system which needs 2 times as much equipment.
3. Lower rated arresters and lower rated BIL required.

Figure 1-12 shows the types of distribution circuits.

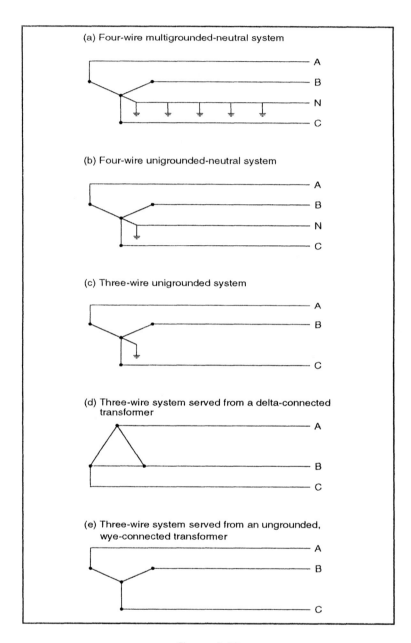

Figure 1-12.

The second most popular type of grounding for a distribution system is the 3-wire delta system shown in Figure 1-12(d). These systems are generally older and lower in voltage than the 4-wire multigrounded type. They are also very popular on industrial power systems. Although still in use, they are not being actively expanded since they lack some of the advantages shown for the 4-wire multigrounded design. Some of the advantages of a 3-wire delta are:

1. Better phase balancing
2. Lower energy into fault
3. Less EMF.

The three other designs shown in Figure 1-12 (12(b), 12(c), 12(e)) are all systems presently in use but only on a very limited basis. Four-wire unigrounded systems (Figure 1-12 (b)) are systems where the primary neutral conductor is insulated at all points except at the source. The neutral conductor in these systems is connected to the neutral point of the source transformer windings and to ground. Distribution transformers usually are connected between phase and neutral conductors with the surge arrester connected between phase and ground. Some four-wire unigrounded systems use an arrester between the neutral conductor and ground. A spark gap may also be used at the distribution transformer between its secondary neutral and arrester ground to provide better surge protection to the transformer windings. The principal advantage of four-wire unigrounded systems is the greater ground relaying sensitivity which can be obtained in comparison to multigrounded systems.

On three-phase three-wire unigrounded primary distribution circuits (see Figure 1-12(c)), single-phase distribution transformers are connected phase-to-phase. The connection of three single-phase distribution transformers or of three-phase distribution transformers is usually delta-ground wye or delta-delta. The floating wye-delta or T-T connections also can be used. The grounded wye-delta connection is generally not used because it acts like a grounding transformer.

LOAD AND FAULT CHARACTERISTICS*

Typical Feeder Load Characteristics

While many distribution engineers are aware of the maximum constraints on system components (e.g., 3-phase underground equipment is generally rated 600 amperes and single-phase underground equipment is generally rated 200 amperes maximum), some of the average characteristics are somewhat less known. Figure 1-13, shown below, illustrates a system design which might be considered fairly typical or average. Obviously, many systems have longer lines (up to 30 miles), higher voltages (up to 34.5 kV), larger fuses (up to 200 amperes), and higher short circuit levels approaching 30 kA.

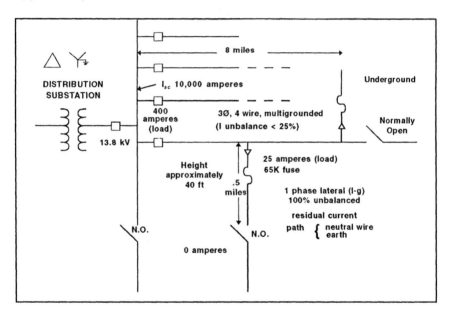

Figure 1-13. Typical Distribution System

---

* These characteristics are developed for 4-wire, multigrounded systems only.

Fault Currents

Fault levels are easily calculated using standard computer programs available to almost all distribution engineers. A good calculation of these levels and a proper understanding of what they mean is very important because fault levels are used to determine protective coordination, switchgear duty, etc. The largest concern when making these calculations is whether to use bolted fault conditions ($Zf = 0$) or a higher fault impedance to reflect the so-called worst case scenario. Studies have shown that there are two fairly distinct types of faults: high impedance and low impedance. If the fault arcs to the neutral, the fault can be classified as "low impedance" reflecting a fault impedance of less than two ohms and usually somewhat less. If, on the other hand, the phase wire lands on the ground and does not contact the neutral conductor, then the fault impedance is very high (usually 300 ohms or more) and the current produced by them is very low, too low to ever be seen by conventional protection practices. Based on the studies described below, experience would indicate that coordination studies using bolted fault conditions is probably the best approach.

Low Impedance Faults. Coordination of protective devices on a distribution system is greatly affected by the short circuit levels on the system. Fault levels are known to vary with distance as illustrated in Figure 1-14. The available short circuit current at the substation is shown to be about 6500 amps, but could be as high as 30 or 40 kA in larger substations near metropolitan areas. The average level reported by utilities at the substation was approximately 10,000 amperes. Many utilities subscribe to the concept that faults have an impedance which consequently limits fault current values to levels considerably less than the calculated bolted fault magnitudes. To compensate for this during coordination studies, some utilities add fault impedance to their short circuit calculations on the order of 5 to 50 ohms. This methodology, while conservative, not only limits the reach of many devices, but as such may cause miscoordination if the actual fault impedance is different from the one used in calculations. There is no question that some faults are "high impedance" faults and as such limit fault current to levels of approximately 50 amperes or less (i.e., hundreds of ohms fault impedance).

Two studies provide the raw data for this discussion. The first is work performed on the Virginia Power 34.5 kV system in 1977 and the

second is edited from work completed for EPRI in 1983 which consisted
of the instrumentation of 50 feeders in thirteen utilities.

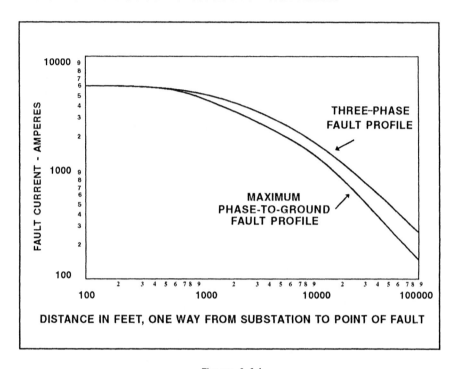

Figure 1-14

In those cases studied where the location of the fault occurrence was
known, it was possible for Virginia Power to calculate the magnitude of
fault current and compare this with the measured values. On the 34.5 kV
circuits, Virginia Power made every effort to include all the resistance and
reactances of equipment, but did not put in any fault resistance. Figure
1-15 indicates the relative accuracy of the calculated values as compared
with the actual measured values. As can be seen, almost 50% of the
calculated values were within ± 5% of the actual implying that there was
little, if any, fault impedance. Most of the error is where the actual
currents were less than the calculated, as would be expected.

In the late 70's, a large 4 year fault study was performed. Fifty
feeders at various voltage levels were chosen to be instrumented at
thirteen utilities.

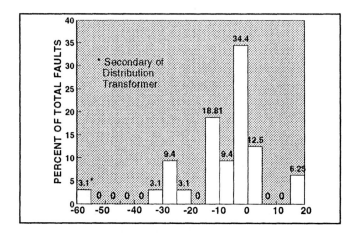

Figure 1-15. Percent Differences of Actual Short Circuit
Currents from Calculated Values

Figure 1-16 indicates the relationship between utility-calculated fault current levels and actual recorded fault currents. In cases where calculated and recorded fault currents are identical, the plotted data point will fall on the 45° line.

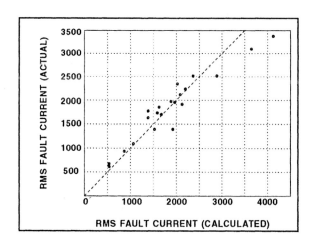

Figure 1-16. Comparison of Actual vs.
Calculated Fault Current Magnitudes

In general, utility calculation of fault currents were extremely close to recorded values. No fault resistance was assumed in the utility calculations, which proved to be a valid assumption, since essentially no fault resistance was seen in the recorded faults. The average utility operated its feeders at approximately 7% above nominal voltage rating, with one recorded instance of a 17% steady-state overvoltage noted. A small amount of fault resistance, on the order of 2 ohms, would explain those faults that fall below the 45° line. Points falling above the 45° line reflect the operating voltage increase above normal or simply the slight inaccuracy of sequence component calculations.

High Impedance Faults. While there have been many high impedance fault studies, many of these do not seem to deal directly with magnitudes of currents during these tests. The information shown on the following page in Table 1-3 illustrates some of these values and indicates the low levels of currents seen during this type of fault. Typically, tests have shown that faults to surfaces such as asphalt, grass, and gravel are less than 50 amps. Faults to concrete, especially wet reinforced concrete, can be of the order of 200 amperes.

Table 1-3. High Impedance Fault Test Summary		
Location	Date	Tests
1. TESCO-Handley	3/78	Normal system switching events were performed: capacitor bank feeder tie breaker operations, large motor start.
2. RG&E	7/11/78	Twelve fault tests consisting of energizing conductor on ground and lowering energized conductor to ground. Conductor laid on dry sod, dry and wet macadam. Arcing faults of 10-50 A on sod. No fault current on dry macadam, slight smoke on wet macadam.
3. TESCO-Handley	8/23/78	Two fault tests on soil each with about 20 A current. Normal switching events were also performed: capacitor bank, feeder tie, and breaker operations.
4. Gulf States	9/20/78	Normal switching events were performed: capacitor bank and feeder tie operations.
5. TESCO-White	2/2/79	Five fault tests with lateral lowered on very wet and partially frozen oil and conductor lowered to touch a grounded metal post. Approximately 20-40 A currents for faults on soil and 50-60 A faults to the post. Arcing occurred in each case.

Table 1-3. High Impedance Fault Test Summary - Con't		
Location	Date	Tests
6. TESCO-White	5/15/79	At White: four fault tests on relatively dry ground. Arcing occurred in each case with about 30 A current. At Handley: normal system data recorded.
7. RG&E	6/1/79	Thirteen faults on dry sod, reinforced and non-reinforced concrete, wet and dry macadam. Several tests with covered conductor produced small fault current on sod. Small fault current on non-reinforced concrete, higher magnitude on reinforced concrete. Little or no fault current on macadam. Subsequent tests on sod produced 50 A currents. Normal system was monitored as were switching events: capacitor bank and load tap changer operations.
8. TESCO-Sand Hills	6/21/79	Eight faults on dry and wetted sand and on a mesquite bush. Less than 1 A current on sand, 30-40 A current on the mesquite bush (due to deep roots).
9. PNM	7/2-3/79	7/2: Instrumentation located at a regulator bank 6 miles from substation. Faults at 1/8, 6, 11, 15, 18 miles from regulator 7/3: Instrumentation at substation, faults at 6, 12, 17 miles. Fault current 10-30 A, faults staged with capacitor in and out and regulator in and out.
10. TESCO-Randol Mill	2/2/80	Eleven fault tests on dry cement and very wet soil. Capacitor bank and load tap changer operations. One fault blew the fuse. Faults on cement had little current, on soil had high magnitude.

Ground Fault Current Flow. Good grounding has been credited with alleviating a variety of utility ills including line flashover, stray voltage and reduced transformer failure rates. Utilities try to maintain 25 ohms or less when placing ground rods but in many utilities obtaining such a low value is just about impossible, regardless of the number of ground rods, due to poor soil conditions. Several utilities have major programs to reduce ground rod impedance. Shown below is a lattice diagram illustrating the flow of ground fault currents for a line-to-ground fault at the end of the feeder. The ground rod impedance used for this calculation was 25 ohms at each pole. It is interesting to note that if the

25

footing resistances are increased, the ground current will decrease, i.e., more current will flow in the neutral and shield wires.

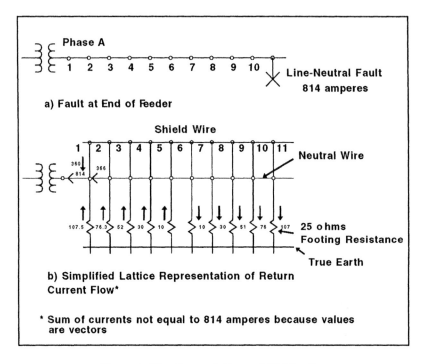

Figure 1-17. Ground Current Distribution

Inrush

Inrush current on a distribution system affects virtually all overcurrent protective device selection and settings. Although there are no comprehensive standards to define inrush, the industry did perform a large study in an attempt to characterize this condition. Figure 1-18 shows a distribution of inrush current magnitudes found in this study. The parameters chosen of peak magnitude, offset ratio, and decay time constant were used primarily because of their importance in fault current characterization for circuit breaker applications. Some general observations on inrush currents are as follows:

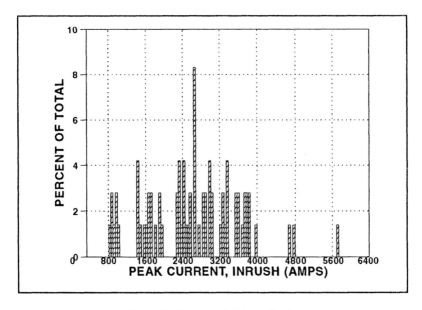

Figure 1-18. Inrush Magnitudes

- The peak instantaneous inrush current averaged 2448 amperes, with a maximum inrush current of 5700 amperes seen in one instance.
- The rms magnitude of the first half-cycle of inrush current was, on average, 2.55 times the rms value of current over the first four cycles. A peak offset ratio of 5.3 was seen in one case.
- The average decay time constant was 53.9 milliseconds (about 3 cycles), indicating that the inrush transient is practically nonexistent after 9 or 10 cycles.
- Average fault current magnitude and average inrush magnitude were approximately the same.

QUESTIONS

1. The most common distribution voltage is 34.5 kV, 2.4 kV or 13.8 kV?

2. Why does short circuit level drop off so fast at relatively short distances from the substation?

3. Short circuit levels should be calculated from the (OA, FA, FOA) rating of a triple rated substation transformer?

4. What are some of the concerns associated with long feeders?

5. Laterals on a typical distribution system are usually fused (true or false)?

6. Name one advantage and one disadvantage of a secondary network system?

7. A typical load level for a distribution feeder is _____ amperes and a typical fuse size for that load is about _____. Why should the lateral fuse size be higher than the load?

8. The most common type of grounding on a distribution system is _____.

9. High impedance faults refer to low current faults at the end of the feeder (true or false)? Explain!

10. Average inrush current levels are not a problem for relay engineers because relay settings are based on fault levels which are much higher in magnitude (true or false)?

2

TRANSFORMERS
AND REGULATORS

TRANSFORMERS

Introduction

The distribution system is unique in that it uses not only many types of transformers but also many different transformer connections. This by itself can make an otherwise simple concept seem more confusing than it should be.

The purpose of this chapter is to present some of the major concepts that concern the distribution systems engineer and operator. Things like types of transformers, loading of transformers, construction and rating are discussed as they are applied to the distribution system. Also, those special transformers, like autotransformers and grounding transformers, are presented since they, too, are common to the distribution system.

Transformer Model

The basic transformer consists of two windings, which are electrically insulated from each other and wound around a laminated core of magnetic sheet steel, as shown in Figure 2-1. Energy is transferred from the source E1, or primary winding, via magnetic flux to the secondary winding. Assuming no transformer losses, power from the source, P1, must equal power to the load, P2. The secondary voltage, E2, is a function of the turns ratio of the transformer, i.e., $E2 = N2/N1*E1$, where N1 is the number of turns in the primary winding and N2 is the number of turns in the secondary winding. Of course this means that as the number of turns on the secondary decreases relative to the primary, as in the case of a step-down transformer, the voltage goes down and the current must go up.

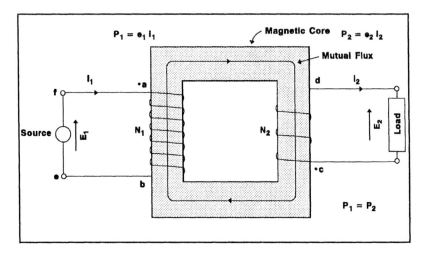

Figure 2-1. Basic for a Lossless Transformer Construction and Relations

The primary and secondary windings are coupled by the transformer entirely through mutual flux, i.e., there is no direct electrical connection since both windings are insulated. As shown in Figure 2-2, not all the flux links the two windings. Some of it "leaks", hence its designation as "leakage flux". It follows that as the load increases and the currents go up so does the leakage flux and hence so do any losses associated with this leakage.

The equivalent circuit for a two-winding transformer is shown in Figure 2-3. As can be seen, there are essentially the following two impedances associated with a transformer.

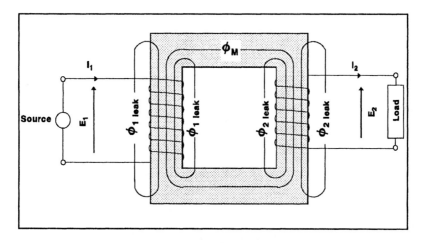

Figure 2-2. Flux Relations of Transformer

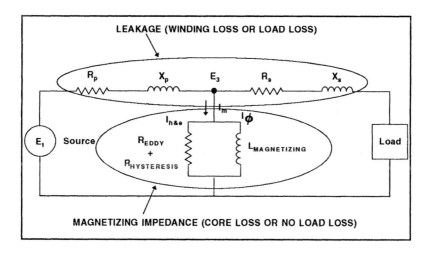

Figure 2-3. Transformer Equivalent

Leakage Impedance. This impedance is composed of Rp+Xp+Rs+Xs or the winding resistances and leakage reactances of the primary and secondary windings. This impedance is also called the "through impedance" and is the impedance used for the calculation of short circuit

31

current and load losses. The impedance referred to on the nameplate of a transformer is this impedance e.g., the impedance of a substation transformer is approximately 10%. This nameplate impedance is associated only with the leakage impedance (also called the winding impedance).

Magnetizing Impedance. When a transformer is energized, it takes a small amount of energy to simply magnetize the iron core (make it hum). The magnetizing branch of the equivalent circuit represents that part of the transformer that consumes energy even under no load and consists of losses due to hysteresis and eddy currents as well as the inductive reactance required to produce the mutual flux. This (magnetizing) impedance is generally quite high and in most cases is neglected since so little current goes through this branch. The nameplate of a transformer normally makes no reference to this impedance since it is rarely used except for transformer loss calculations.

The current through this impedance varies only slightly with load since E3 is relatively constant under variable load conditions. The magnetizing branch results in losses which are relatively constant since both current in this branch and voltage across this branch are essentially constant. Losses associated with this branch are called "no load" losses since they occur whether there is load on the transformer or not. An equivalent circuit for "no load" is shown in Figure 2-4.

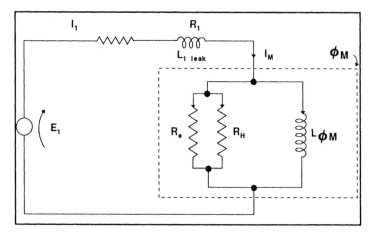

Figure 2-4. Primary Circuit of a Transformer Under No Load

Autotransformers

The autotransformer is a special type of transformer that is used in various types of distribution applications, primarily where the voltage transformation is small and cost is a major consideration. The equivalent circuit for a step-down autotransformer is show below. As can be seen, the thing which distinguishes an autotransformer from a standard two-winding unit is the direct electrical connection between the primary and secondary. While this connection has advantages in the areas of voltage regulation and energy transfer, it has drawbacks concerning high short circuit currents (because the transformer leakage impedance is usually less than 5%) and the fact that the primary is not completely insulated from the secondary.

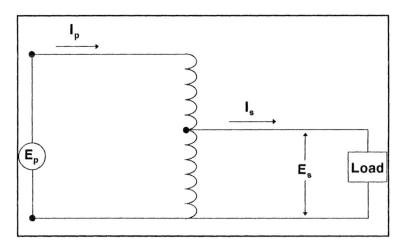

Figure 2-5. Circuit of Step-Down Autotransformer

Probably the best way to demonstrate the advantage of an autotransformer is with a hypothetical example. Suppose you purchased a 2-winding, single-phase, 25 kVA distribution transformer rated 7200 volts on the primary and 240 volts on the secondary, as shown in Figure 2-6a. We know that the power in (V_p*I_p) must equal the power out (V_s*I_s) and both should equal 25 kVA.

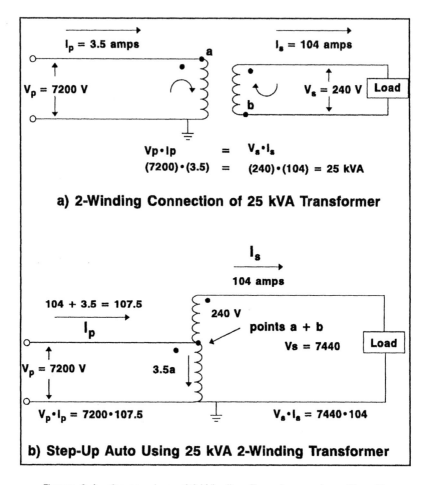

Figure 2-6. *Conversion of 2 Winding Transformer to a Step-Up Autotransformer*

If we reconnect the transformer so that point "a" of the primary is physically attached to point "b" and the load is connected between the phase wire and ground, we now have a step-up autotransformer (see Figure 2-6b). The secondary voltage is now higher than the primary by 240 volts, i.e., it's 7200 volts plus 240 volts or 7440 volts. The magnetic power transfer of the transformer is exactly the same as it was since the currents through the windings are the same as they were (as are the voltages across each winding). The voltage across the load and the current from the source, however, have both increased dramatically to about 774 kVA. This means that the power transfer capability of this transformer has been increased by over 30 times, which is mainly the result of the direct electrical transfer of energy now taking place.

Construction

There are two types of transformer construction commonly referred to in the utility. These are referred to as "core type" and "shell type" and are shown in Figure 2-7. The obvious difference in these two types of construction is that the windings in the shell type construction are surrounded by core (iron) material. Some manufacturers will make the claim that this makes the transformer inherently more capable of handling the high mechanical forces imposed on the transformer during a short circuit. While in theory this may be true, in actuality core type transformers produced by quality manufacturers have reliability rates as good or even better than most manufacturers of shell type.

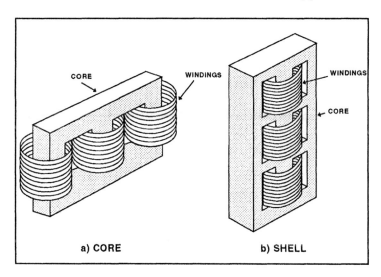

Figure 2-7. Core and Shell Type Transformer Construction

Types of Transformers

Distribution Substation Transformers.   Distribution substation transformers come in such a wide variety of ratings that it would be impossible to cover them all.   Figure 2-8 illustrates what might be considered a typical substation size and layout.

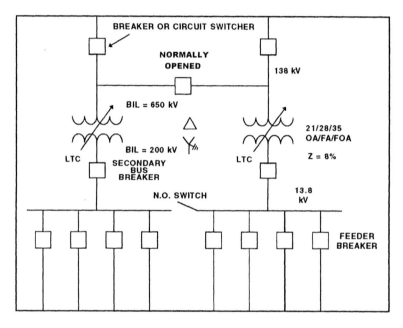

Figure 2-8. Typical Distribution Substation

As can be seen, this 2-winding transformer has a 138 kV to 13.8 kV voltage transformation with an 8% transformer impedance.   Some substations use autotransformers which usually have less than half that impedance.   Some typical characteristics (not all inclusive) that the distribution engineer might find in a utility system substation are as follows:

High Voltage Winding - 230 kV to 34.5 kV
Low Voltage Winding - 34.5 kV to 4160 volts
Size in MVA (OA) - 2.5 to 75 MVA per transformer
Transformer Impedance - 5% to 12%

Number of Transformers in Substation - 1 to 4
Loading - OA, OA/FA, OA/FA/FOA, OA/FA/FA
High Side Protection - Circuit Switchers, Breakers, Fuses
Relay Protection - Overcurrent, Differential, Under-Frequency
Feeder Protection - Breakers, Reclosers

Most distribution substation transformers larger than 5 or 10 MVA are protected by breakers or circuit switchers and use both differential and overcurrent relays. These transformers are usually regulated by LTC (load tap changers) which adjusts voltage level depending on loading conditions. The secondary bus is usually split (normally open) to reduce the number of short circuits seen by each transformer as well as to reduce the short circuit level of the system. Keeping this bus tie open also reduces any circulation currents that might take place should the transformer impedances not be exactly matched.

Loading of the transformer is usually kept within its OA rating (passive convection), with the FA (fans) and FOA (fans and oil pumps) ratings utilized for emergency conditions. Both the FA and FOA ratings increase the loading by about 33% each in transformers rated 10 MVA and greater.

Two calculations the reader should be familiar with in regard to a distribution substation transformer are load current and short circuit current. They are as follows for the transformers shown in Equations 2-1 through 2-3.

Calculation of Full Load Current:

$$21 \text{ MVA} = 7 \text{ MVA per phase}$$
$$13.8 \text{ kV} = 7.97 \text{ kV line-to-ground}$$
$$\text{Load amps (an OA rating)} = \frac{7 \text{ MVA}}{7.97 \text{ kV}} = 878.3 \text{ amperes} \qquad 2\text{-}1$$
$$\text{FA rating} = 878 \times 1.33 = 1168 \text{ amperes}$$
$$\text{FOA rating} = 878 \times 1.66 = 1457 \text{ amperes}$$

Calculation of Short Circuit (line-ground):

$$I_{SC} = \frac{\text{LOAD AMPS (OA*)}}{\text{Per Unit Transformer Impedance}}$$

2-2

$$= \frac{878.3}{.08} = 10,980 \text{ amperes}$$

Alternate method:

$$I_{SC} = \frac{\text{Voltage (l-g)}}{\text{Transformer Z}}$$

$$\text{Transformer Z} = Z_{(p.u.)} * Z_{BASE} = Z_T$$

$$Z_{BASE} = \frac{kV^2}{MVA} = \frac{13.8^2}{21} = 9.07 \text{ ohms}$$

2-3

$$Z_T = .08 * 9.07 = .7256 \text{ ohms}$$

$$\therefore I_{SC} = \frac{7.97 \text{ kV}}{.7256 \text{ ohms}} \approx 10,980 \text{ amperes}$$

*Only the OA rating should be used to calculate short circuit currents.

Distribution Transformers. The transformer that connects the high voltage primary system (4.16 kV to 34.5 kV) to the customer (at 480 volts and below) is usually referred to as a "distribution transformer". The variety of ratings, transformer connections, tank configurations, protection schemes, and loading practices are much too diverse to be covered by a single chapter so only some of the more major areas can be discussed.

These transformers can be either single-phase or three-phase and range in size from about 5 kVA to 500 kVA. Table 2-1 shows some of the standard kVAs and voltages for these units. Pad mounted transformers, used for underground service come in about the same ratings and sizes with the exception that 3-phase padmounted units can be rated as high as 2500 kVA. Impedances of these 2-winding transformers are generally quite low, ranging from about 2% for units less than 50 kVA to about 4% for units greater than 100 kVA.

| Table 2-1. Standard Transformer KVAs and Voltages ||||||
| KVAs || High Voltages || Low Voltages ||
Single-Phase	Three-Phase	Single-Phase	Three-Phase	Single-Phase	Three-Phase
5	30	2400/4160Y	2400	120/240	208Y/120
10	45	4800/8320Y	4160Y/2400	240/480	240
15	75	2400/4160Y	4160Y	2400	480Y/277
25	$112^1/_2$	4800Y/8320YX	4800	2520	240X480
$37^1/_2$	150	7200/12,470Y	8320Y/4800	4800	2400
50	225	12,470GrdY/7200	8320Y	5040	
					4160Y/2400
75	300	7620/13,200Y	7200	6900	4800
100	500	13,200GrdY/7620	12,000	7200	12,470Y/7200
167		12,000	12,470Y/7200	7560	13,200Y/7620
250		13,200/22,860GrdY	12,470Y	7980	
333		13,200	13,200Y/7620		
500		13,800/23,900GrdY	13,200Y		
		13,800	13,200		
		14,400/24,940GrdY	13,800		
		22,900	22,900		
		34,400	34,400		
		43,800	43,800		
		67,000	67,000		

Pole type, as well as underground transformers, can be either single-phase, or three-phase depending on the requirements of the load and the configuration of the primary supply. Figure 2-9 shows some of the more common pole type configurations. As can be seen, single-phase transformers can have one or two primary bushings. If the primary system was delta connected (no neutral wire) then the primary of the transformer would have to be connected phase-phase requiring two bushings. If the system was a 4-wire multigrounded system, then single-phase loads would normally be connected phase-to-neutral (ground) and only one primary bushing would be required (some utilities prefer 2 primary bushings even on these systems for reasons of flexibility).

Some of the more common high-voltage connections seen at the transformer are shown in Figure 2-10. As can be seen the winding configuration affects the way the voltage rating is applied. This is primarily because grounded wye transformers may have tapered insulation (less insulation near the neutral) and as such cannot be applied on an ungrounded system.

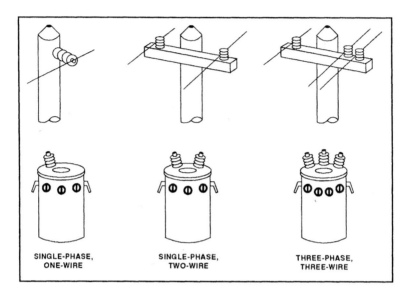

Figure 2-9. Pole-Type Transformer Connections

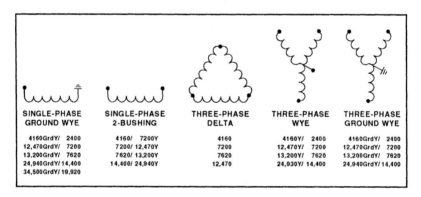

Figure 2-10. High Voltage Connections

Distribution transformers are assigned kVA ratings which indicate the continuous load the transformers carry and not exceed the specified temperature rise of either $55^{\circ}C$ or $65^{\circ}C$. In service, a distribution transformer is rarely loaded continuously at its rated kVA and typically goes through a daily load cycle. Because the transformer has a relatively long thermal time constant, i.e., the oil temperature goes up slowly relative to the load change, it is possible to load the transformer beyond its rating without serious effect on the life of the unit. Table 2-2 shows a typical loading guide. Many utilities have successfully used even more severe loading guides for their smaller single-phase distribution transformers with no apparent problem with life expectancy.

Table 2-2. Permissible Daily Short-Time Transformer Loading Based on Normal Life Expectancy			
Period of Increased Loading, Hours	Maximum Load in Per Unit of Transformer Rating		
	Average Initial Load in Per Unit of Transformer Rating		
	0.90	0.70	0.50
0.5	1.59	1.77	1.89
1.0	1.40	1.54	1.60
2.0	1.24	1.33	1.37
4.0	1.12	1.17	1.19
8.0	1.06	1.08	1.08

Distribution transformers are manufactured in a variety of ways from as simple as a 2-winding transformer with no protection to a "self-protected" (CSP) transformer illustrated in Figure 2-11. The self protected distribution transformers arrive to the user equipped with a primary surge arrester, an internal high voltage oil immersed expulsion (weaklink) fuse and a low voltage circuit breaker that provides protection from some secondary faults and severe overloads.

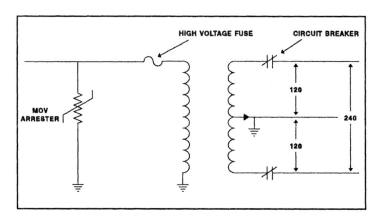

Figure 2-11. "Self-Protected" Distribution Transformer

The low voltage circuit breaker trips on the basis of temperature. This tripping temperature is caused by the combination of surrounding oil and the current flow in the bi-metallic sensor. As an example, a 25 kVA transformer having 75% initial load, and a 35C ambient would initiate an overload "light" and warning after approximately 2 hours of 2-per-unit overload and a secondary breaker trip in approximately 2 1/2 hours at the same overload level. Finally, many utilities do not use secondary breakers because they find them more of a nuisance than a help.

Network Transformers. Network transformers are large (300 kVA to 2500 kVA), liquid filled, three-phase transformers used to supply secondary network systems (see Figure 2-12). The high voltage windings are generally anything from 4160 volts to 34.5 kV and the secondary voltage rating is usually 216GrdY/125 or 480GrdY/277 volts. New systems and virtually all spot networks, used for highly concentrated loads such as commercial buildings, use the higher secondary voltage rating. Since the units are usually installed under streets, they must be sealed and corrosion resistant.

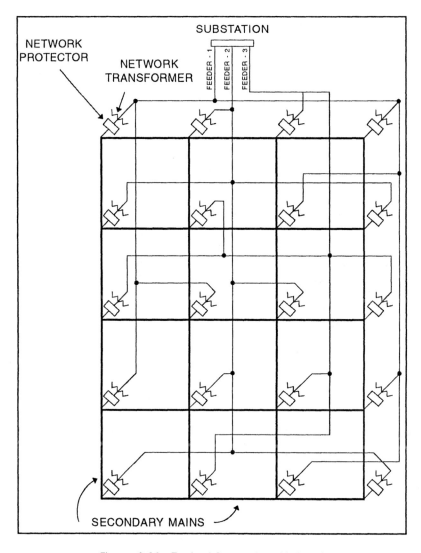

Figure 2-12. Typical Secondary Network

The secondary network vault usually contains three fundamental components as shown in Figure 2-13:

1. **High Voltage Switch.** This switch allows the transformer to be connected and disconnected to the primary feeder. It does not have

load break capability and can be opened and closed only when the network protector is opened.

2. **Network Transformer.** These 3-phase transformers are liquid (PCB or silicon) filled, and connected either delta/GrdY or GrdY/GrdY. Units 1000 kVA or smaller generally have an impedance of about 5% whereas the larger units, 1500 kVA and above, have an impedance closer to about 7%.

3. **Network Protector.** The network protector consists of an air circuit breaker and relays that operate for reverse power flow (e.g., primary fault). The network protector also has a fuse whose purpose it is to backup the network protector breaker as well as provide some thermal overload capability to the transformer. This fuse is not sized to provide protection against secondary faults. Secondary faults are generally considered to be "self-clearing", i.e., they burn themselves clear or are cleared by fuses at the cable's junctions called "current limiters". Because many faults at that level have been found not to burn themselves clear, 480 volt network systems pretty much require "current limiters".

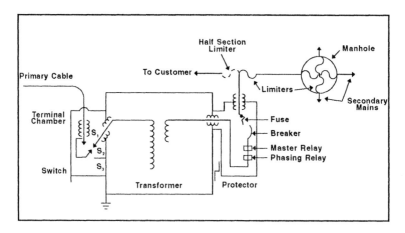

Figure 2-13

Dry Type Transformers. Dry type transformers are transformers whose insulation medium is solid (glass tape, glass silicon, porcelain, etc.) and not a liquid like oil. Dry type transformers are primarily used where safety is a major concern e.g., transformers in buildings in close proximity to people. With the elimination of askarel (PCB's) as an insulation medium for transformers, for environmental reasons, the importance of "dry type" transformers significantly increased. The kV and kVA ratings of dry type transformers are very similar to oil filled transformers. Standard kVA ratings range from below 5 kVA up to 20 MVA with voltage ratings from less than 2400 volts up to 230 kV.

Dry type transformers are not considered as rugged as oil filled transformers of the same rating. To insure reliability most users have found it prudent to never exceed the loading recommendations of the manufacturers. This is somewhat in contrast to the loading practices of some utilities where overloading beyond the standards of liquid filled units is somewhat routine and has not produced excessive failure rates. Also, a concern with dry type transformers is their lower dielectric strength. Table 2-3, shown below, illustrates that a 13.8GrdY/7970, dry type transformer has BIL of 60 kV. This is only 63% of an equivalent oil filled unit which has a BIL of at least 95 kV. Special dry type transformers are made with increased BIL.

Table 2-3. Insulation Levels for Dry Type Transformers			
Nominal Equipment Voltage (1)	BIL (kV) (2)	Nominal Equipment Voltage (1)	BIL (kV) (2)
120-1200 1200GrdY/693	10	12,000 13,800 13,800GrdY/7970	60
2520 4360GrdY/2520	20	18,000 22,860GrdY/13,200	95
4160 7200 8720GrdY/5040	30	23,000 24,940GrdY/14,400	110
8320	45	27,600 34,500GrdY/19,920	125
		34,500	150

Secondary Faults

The calculation of bolted fault currents on the secondary of the distribution transformer is sometimes necessary in order to select service entrance equipment with adequate interrupting rating as well as to determine overcurrent coordination of secondary protective devices. Since the fault can occur across either the 120 volt or 240 volt circuits, this calculation can sometimes get a little confusing. The equations shown below can be used for this purpose as well as to calculate secondary faults on the secondary cable itself.

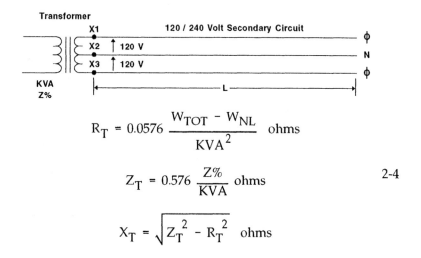

$$R_T = 0.0576 \frac{W_{TOT} - W_{NL}}{KVA^2} \text{ ohms}$$

$$Z_T = 0.576 \frac{Z\%}{KVA} \text{ ohms} \qquad 2\text{-}4$$

$$X_T = \sqrt{Z_T^2 - R_T^2} \text{ ohms}$$

KVA = Transformer nameplate rating in kVA
W_{TOT} = Transformer total losses at full load in watts
W_{NL} = Transformer no load losses in watts
Z% = Transformer nameplate impedance in percent
R_T = Transformer resistance in ohms at secondary terminals X1-X3
Z_T = Transformer leakage impedance in ohms at secondary terminals X1-X3
L = Circuit length from the transformer secondary terminals to fault in feet

$$I_{240} = \frac{240}{\sqrt{\left(R_T + \frac{R_S L}{1000}\right)^2 + \left(X_T + \frac{X_S L}{1000}\right)^2}} \text{ amps}$$

2-5

$$I_{120} = \frac{120}{\sqrt{\left(.375 \ R_T + \frac{R_{S1} L}{1000}\right)^2 + \left(.5 \ X_T + \frac{X_{S1} L}{1000}\right)^2}} \text{ amps}$$

I_{240} = Available current for a bolted 240 volt (phase-to-phase) fault in amperes rms symmetrical

I_{120} = Available current for a bolted 120 volt (phase-to-neutral)fault in amperes rms symmetrical

R_S = Resistance of secondary circuit for a 240 volt fault in ohms per 1000 feet

R_{S1} = Resistance of secondary circuit for a 120 volt fault in ohms per 1000 feet

X_S = Reactance of secondary circuit for a 240 volt fault in ohms per 1000 feet

S_{S1} = Reactance of secondary circuit for a 120 volt fault in ohms per 1000 feet.

DISTRIBUTION TRANSFORMER CONNECTIONS

The combination of lighting and large 3∅ power loads, brings up the question as to "what is the best transformer connection?" This is not an easy question since there is such a wide variety of connections and considerations therein. Some of the more common areas of interest are described as follows:

3∅ Secondary Voltage

An important consideration for supplying 3∅ load is whether it should be connected to a delta or wye connected secondary (see Figure 2-14).

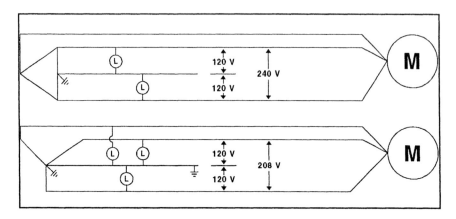

Figure 2-14. Secondary Loading of Delta and Wye Connection

It will be noted in the delta connection that all of the single-phase load is concentrated on one phase, whereas on the wye connection all the single-phase (lights and small appliance) load is distributed on each of the three phases.

It is evident that if a large number of single-phase loads are to be served from the bank, the delta secondary connection is less suitable, because of the unbalanced loading. On the other hand, if the single-phase load is small, as compared with the polyphase load, the delta connection is preferable.

In urban distribution systems, particularly in heavy commercial areas, the trend is toward the four-wire, three-phase, 120/208Y volt system. In rural distribution, where the need for three-phase secondary is limited to isolated loads, the four-wire, three-phase, 120/240 volt delta system is perhaps more common.

In selecting a secondary transformer connection, the performance characteristics of motors and other load devices at other than rated voltage, must be considered. For example, with the delta connection, a 120 volt standard lamp and a 220 volt three-phase motor will be applied at their name plate rating; whereas, with the wye connection, the 120 volt lamp will have name plate voltage applied but only 208 volts will be available for power purposes. Unless 208 volt motors are explicitly specified, standard 220 volt motors will not perform as might be expected.

Wye-Delta

This bank can have the neutral grounded or isolated. With the neutral grounded the bank may be operated as open wye-open delta at reduced capacity if one phase in the primary is lost but it also acts as a grounding bank for the system and as such will replace the main source in feeding short circuit current to a line-to-ground fault after the protective devices have opened. To avert this and possible transformer burnout, it is recommended that a three-phase wye-delta bank be ungrounded.

With transformer connected wye-delta and the neutral isolated, if one phase of the primary is opened by a fuse blowing or for any other reason, the ungrounded bank ceases to operate as a three-phase bank and is unable to carry a three-phase load. It is therefore a safe practice to have three element protectors on all three-phase motors fed from this bank.

Delta-Wye

This bank is similar to the wye-delta bank in many of its characteristics and it may also have the neutral grounded or isolated. With the neutral grounded the bank is more flexible in the sense that it may consist of units of different rating as determined by 1/3 of three-phase load plus whatever single-phase load is connected to it. Only 120/208 volt service will be available from this bank, compared to 120/240 volt three-wire service that may be obtained from wye-delta banks. If the neutral is isolated and is not carried through as a fourth wire, the bank will not be expected to supply 120 volt loads.

This type of connection, i.e., delta-wye with the neutral grounded is the most popular connection in distribution substation banks where transformation may be 69 kV to 7200/12470GrdY or 2400/4160Grd wye.

Delta-Delta

This type of connection should be used only when the percentage impedance and the ratio in all units are identical. Even a slight difference in the voltage ratio among the units may cause large circulating currents to flow in the windings which can reach dangerous proportions and heat up the transformers even under light loads. A different setting of taps is in its effect equivalent to units of different transformer ratios; therefore, they should have not only the same voltage ratio but also the same voltage tap setting.

This bank connection will provide the possibility of serving a three-phase load with only two transformers connected open delta. This is an emergency measure at which the total bank output will be 57.7% of the original bank capacity.

Wye-Wye

The most common connection that may be considered under this heading is the neutral grounded on both sides of the bank. In a three-phase distribution system with a multigrounded neutral, the grounding of the primary and secondary neutral of a service bank appears to be a good solution for feeding single-phase and three-phase loads with a 120/208 volt supply. One reservation to this type connection is the necessity of using 208 volt motors instead of standard 220 volt motors, unless the 120/208 volt system is a part of a secondary network system, with very close voltage regulation.

This type of connection is sometimes used on 14.4/24.9 kV systems where experience has shown that the ungrounded wye-delta bank is cause for critical overvoltages when a phase is de-energized.

The wye-wye grounded neutral bank may be made up of transformers selected as the load on each phase requires without limitation in their characteristics and kVA ratings.

Third harmonic currents will circulate freely in the primary and secondary circuits of the bank and may become a source of telephone interference when telephone lines run close to or parallel to the power circuit.

Open Delta

The use of open delta or "V" connected transformers for a three-phase service is not to be considered as a standard practice and it is recommended only in special cases as listed below. The bank must be operated at reduced capacity and its regulation is different in each leg creating an unbalanced voltage condition on the secondary. Therefore, its use should be considered only under the following conditions:

1. If one of the transformers of a wye-delta bank burns out or must be disconnected, the two remaining transformers can be operated to furnish three-phase load at reduced capacity provided the primary neutral is connected to ground and common neutral wire.

2. If only two phases are available on the primary line and it is considered uneconomical to string the third phase to pick up a three-phase load.

    Transformers connected in open delta or "V" connected are able to carry 86% of their rated capacity. Voltage regulation in each phase will be different for different values of load power factor.

Grounding Banks

An objection to a wye-delta connection is that it tries to act as a grounding bank during a fault at some other point on the system. Figure 2-15 shows how fault currents flow from the substation transformer with a line-to-ground fault at "X". The currents in the step-down wye-delta bank where the high voltage neutral of the transformer bank is isolated, are not affected by the fault at "X".

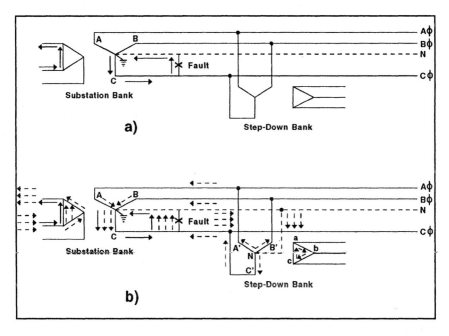

Figure 2-15

Figure 2-15b shows the same system, except that the neutral of the step-down transformer bank is connected to the neutral conductor. With a fault at "X", the same fault current flows from the substation transformer as in Figure 2-15a. These currents are indicated by the solid arrows in Figure 2-15b.

The fault at "X" reduces the voltage across C'N which reduces the voltage across ac, causing circulating currents in abc that appear as currents in A'n, B'N, and C'N. A very low resistance fault at "X" will result in a practical collapse of all voltage across C'N, producing extremely heavy circulating currents in the transformer windings. These currents (dotted arrows in Figure 2-15b) add to the normal load current in the transformer bank and are apt to burn it out.

Summary

A summary of the characteristics of single-phase transformers connected for combined 3Ø and 1Ø loads is shown in Table 2-4.

Table 2-4. Distribution Transformer Connection Guide

	1	2	3	4	5	6	7	8	9
PRIMARY CONNECTION	Y	Y	Δ	Δ	Δ	Y	Y	Y	Y
SECONDARY CONNECTION	Δ	Δ	Δ	Y	Y	Y	Y	Y	Y
1. Suitable for 120 volt single-phase, 208 volt three-phase load.									X
2. Suitable for only 208 volt single-phase or three-phase load.								X	
3. Suitable for 120 / 240 volt load.	X	X	X						
4. Self-protected transformers suitable.							X		X
5. Primary ground fault may cause transformer burnout; fault fed back through bank.		X							
6. Only single phase feed available with one phase open.	X			X	X	X	X	X	X
7. Use overcurrent protection on all leads of three-phase motors.	X			X	X				
8. Suitable for unbalanced or single-phase load.	X	X	X		X		X		X
9. Percentage impedance and ratio may differ in all units.	X	X	X		X			X	X
10. KVA rating of all units may differ.						X			
11. Third harmonic current objectionable.							X		
12. Dangerous third harmonic voltages.									

REGULATORS

Introduction

The voltage regulator is simply a tapped autotransformer and as such works in a similar fashion, to either raise or lower voltage. On a distribution system, regulators can be found in the substation where they are referred to as station or bus regulators, or out on the feeder where they are referred to as supplementary or pole type regulators. Because an autotransformer uses both a direct electrical connection as well as magnetic flux to transfer energy, the regulator nameplate rating in kVA will be much lower than the kVA of load it can regulate. Typically, most of the energy transfer (approximately 90%) of the regulator comes via the direct electrical connection when the regulator is either in the full boost or full buck mode. At nominal tap setting virtually all the energy is transferred by direct electrical connection allowing some regulators to have higher kVA ratings for reduced tap settings.

Figure 2-16, shown below, shows a simple schematic of a single-phase voltage regulator. As can be seen, a typical voltage regulator used out on the feeder (pole mounted), usually has a bucking range equal to the boost range or ±10%. Each tap normally gives a 5/8% voltage change, so there are generally 32 taps.

A typical example of the use of a feeder regulator is shown in Figure 2-17. In this example, the supplementary feeder voltage regulator is located at a point on the feeder where the voltage starts to dip below 119 volts. The voltage standards require a voltage to the residential customer to be between 126 and 114 volts. This voltage is measured at the meter. In order to compensate for the voltage drop from the feeder to meter, i.e., the drop through the transformer and secondary, the feeder voltage must be somewhat higher than 114 volts. Some utilities use a 5 volt drop to estimate this reduction thus requiring a 119 voltage (114V+5V = 119 volts) on the feeder itself. Another point worth noting is that the feeder voltage can be higher than 126 volts as long as the voltage by the time it gets to the customer meter is reduced to 126 or less.

Regulator kVA Rating

Single-Phase. The rating of a single-phase feeder voltage regulator is the product of the rated load amperes and the rated range or regulation in kilovolts. The rated range of a regulator with a ±10% capability is 10%

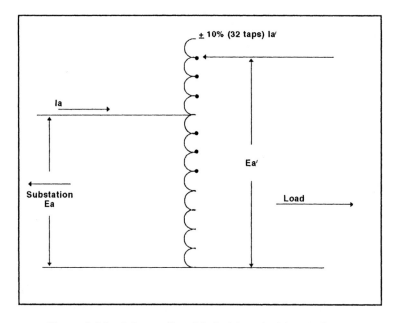

Figure 2-16. Schematic of Pole Mounted Single-Phase Regulator

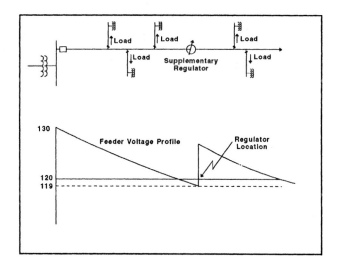

Figure 2-17. Supplementary Feeder Regulation

not 20%. For example, the minimum size single-phase regulator needed to supply 100 amperes on a 7620 volt line-to-ground system (4-wire multigrounded system) would be calculated as follows:

$$1\emptyset \text{ Regulator kVA} = \text{Range} \times \text{System kVA}$$
$$= .10 \times 7620V \times 100 \text{ amps} \qquad 2\text{-}6$$
$$= 76.2 \text{ kVA (minimum)}$$

For a single-phase delta rated 13200 volts, phase-to-phase, supplying a load of 100 amperes, the minimum regulator rating would be:

$$1\emptyset \text{ Regulator kVA} = \text{Range and System kVA}$$
$$= .1 \times 13200 \text{ volts} \times 100 \text{ amps} \qquad 2\text{-}7$$
$$= 132 \text{ kVA (minimum)}$$

Three-phase. Three-phase voltage regulators can be applied either to three-phase, three-wire or three-phase, four-wire circuits. The kVA ratings are determined in the same manner for each type system and the same regulators can be used. As shown in Figure 2-18, the phases are normally wye connected internally. On the four-wire circuits the regulator neutral connection is connected to the fourth neutral wire. On three-wire ungrounded circuits the neutral is normally connected through a surge arrester to ground.

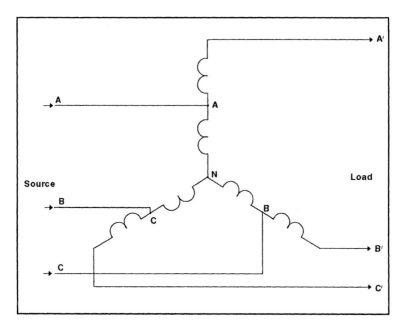

Figure 2-18. Three-Phase Regulator for Use on Three-Phase, Three-Wire or Three-Phase, Four-Wire Circuit

When single-phase regulators are to be used on three-phase circuits, the regulator arrangement for a three-phase four-wire circuit is quite different from that on a three-phase three-wire circuit.

On a three-phase, four-wire circuit, three, single-phase regulators, whose voltage rating is the line to neutral voltage rating of the circuit, are connected in wye, as shown in Figure 2-19.

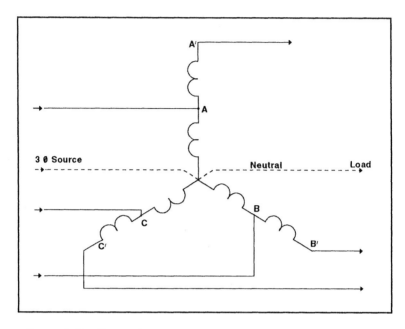

Figure 2-19. Three Single-Phase Regulators Wye Connected on Three-Phase, Four-Wire Circuit

On the three-phase, three-wire circuit the usual connection is to use two regulators, whose voltage rating is the line to line circuit voltage in open delta as shown in Figure 2-20. Two ±10% regulators will provide ±10% regulation on all three phases as shown in the phasor diagram of Figure 2-21.

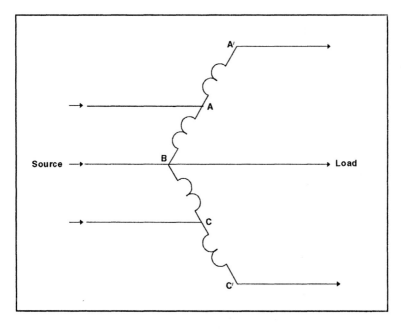

Figure 2-20. Two Single-Phase Regulators Connected in Open-Delta on Three-Phase, Three-Wire Circuit

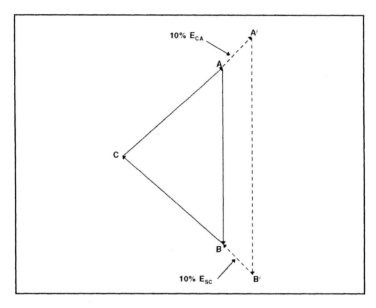

*Figure 2-21. Phasor Diagram Showing Two Single-Phase
Regulators Connected in Open-Delta*

It is a relatively simple matter to determine the kVA rating of a regulator if the necessary ground rules are kept in mind.

1. The kVA rating of a <u>single-phase feeder-voltage regulator</u> is the product of the rated load amperes and the rated range or regulation in kilovolts.
2. The kVA rating of a <u>three-phase feeder-voltage regulator</u> is the product of the rated load amperes and the rated range of regulation in kilovolts multiplied by 1.732.
3. The rated range of regulation of a regulator is the amount that the regulator will raise or lower its rated voltage.

When regulators are being considered for a given circuit, several pertinent facts must be known. These facts are the type of circuit (single-phase or three-phase), the voltage rating of the circuit, the kVA rating of the circuit, and the required amount of voltage correction. In general, the required correction will fall within the range for which standard regulators are designed. Although the basic rules apply for any range, the standard range of plus or minus ten percent will be used for the following examples. From this information, we can determine the rated current to complete the information necessary to calculate the regulator kVA. Let us take several examples to show how the rules work.

Example 1. Three-Phase, Three-Wire or Four-Wire Circuit, with Three-Phase Regulator as Shown in Figure 2-18.

Assume: 7960/13,800Y-circuit, rated 4000 kVA for ±10% range

Rated Load Amperes =

$$\frac{kVA \times 1000}{V_{L-L} \times 1.732} = \frac{4000 \times 1000}{13,800 \times 1.732} = 167 \text{ amperes}$$

Range in Kilovolts =

$$Range \times kV_{L-L} = 0.10 \times 13.8 = 1.38$$

Regulator kVA =

$$167 \times 1.38 \times 1.732 = 400 \text{ kVA}$$

Therefore, the next highest standard rating available will be a three-phase regulator rated 500 kVA, 13,800 volts, 209 amperes, 10% raise and lower.

Example 2. Three-Phase, Four-Wire Circuit, with Three Single-Phase Regulators as Shown in Figure 2-19.

(This is the same as a single-phase regulator on a single-phase circuit.)

Assume: 7960/13,800Y-circuit, rated 4000 kVA with ±10% range

Single-phase regulators would be excited from line to neutral at 7960 volts with the rating calculated as follows:

Load Amperes =

$$167 \text{ (Same as Example 1)}$$

Range in Kilovolts =

$$0.10 \times 7.96 = .796 \text{ kV}$$

Regulator kVA =

$$167 \times .796 = 133 \text{ kVA}$$

Each single-phase regulator should be rated 167 kVA, 7960 volts, and 209 amperes, 10% raise and lower

Example 3. Three-Wire Delta Circuit, with Two Single-Phase Regulators as shown in Figure 2-20.

Assume: 13,800 volt circuit, rated 4000 kVA with ±10% range

Load Amperes =

$$167 \text{ (Same as Example 1)}$$

Range in Kilovolts =

$$0.10 \times 13.8 = 138$$

Regulator kVA =

$$167 \times 1.38 = 230 \text{ kVA}$$

Each single-phase regulator would be rated 276 kVA, 13,800 volts, 200 amperes, 10% raise and lower as a standard unit of the next highest rating to meet application requirements.

For feeder-voltage regulators in general, the actual kVA of regulation required bears the same relation to the circuit kVA as the regulating range bears to the voltage. For example, as the examples show in Case 1 and 2, a 150 kVA three-phase regulator would provide a range of 10 percent in a 1500 kVA feeder. Likewise, a 50 kVA single-phase regulator would provide a range of 10 percent in a 500 kVA single-phase circuit. This is a useful rule-of-thumb that can be used for all normal feeder-regulator application. With the regulator kVA requirement known, it is a simple matter to pick a standard rating equal to requirement or the next higher rating.

QUESTIONS

1. The impedance given on the nameplate of a transformer does not include the magnetizing impedance. Explain.

2. Explain the difference between the load loss and the no load loss of a transformer.

3. Why can an autotransformer be so efficient at carrying load?

4. Calculate the line-to-ground short circuit level for a substation autotransformer rated 20 MVA, 13.8 kV and having a 2% impedance. Name one problem with using this autotransformer.

5. How are network cable faults cleared? And describe the function of a network protector.

6. Why are "dry type" transformers more prone to failure?

7. What is the major objection to the use of a wye-delta distribution transformer connection? Explain.

8. Which transformer connections produce high third harmonic voltages?

9. What is the normal range of a regulator?

10. Determine the three-phase regulator rating to be used on a four-wire distribution system, rated 19.9/34.5 kV, having a ±10% range, and serving a load of 5 MVA.

3

APPLICATION OF CAPACITORS FOR DISTRIBUTION SYSTEMS

FUNDAMENTALS

Early distribution systems consisted of loads that were primarily lighting. Since lighting is a resistive type of load, the power factor of these systems was inherently quite high. Today, however, much of the utilities' load is motor load, which contains reactive power requirements. This low power factor reactive load can cause voltage drop and losses which are largely unnecessary.

With the higher cost of losses resulting from high prices for fossil fuels, utilities have found it beneficial to correct system power factor to almost unity. Also, with the massive mandatory PCB replacement programs placed upon the utilities and sophisticated computer programs becoming available, new concepts to maximize the effectiveness of capacitor placement are commonplace (see Figure 3-1).

Real and Reactive Power

The system shown in Figure 3-2 consists of an ac source connected directly to a lighting load. The effective voltage and current are designated by E and I. In a resistive circuit of this type, the voltage and current would be in phase. A wattmeter connected to the line would give a reading of $P = E \times I$ in watts (real or active power).

To get a better idea of the in phase interaction of E and I in a resistive circuit, the sinusoidal curves for E, I, and P are shown in Figure 3-3 below. The power factor in this example is unity ($\cos \theta = \cos 0^o = 1$).

Since E and I are **effective** values, the peak values are the $\sqrt{2}\ E$ and the $\sqrt{2}\ I$. The power wave is a double frequency (120 Hz) flow consisting of all positive loops where average P = PEAK POWER/2 $= \sqrt{2} \cdot \sqrt{2} \cdot EI/2 = EI$. This indicates that power is positive and always

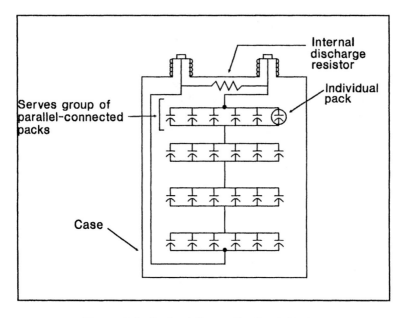

Figure 3-1. Typical Capacitor Bank Design

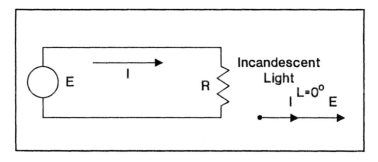

Figure 3-2. Phase Relationship of Resistive Circuit

flowing from the source into the resistor. This is one of the basic properties of active (or real) power.

On the other hand, the circuit below in Figure 3-4 illustrates the concept of reactive power. This circuit is comprised of an ac source feeding an inductance such as a motor load.

65

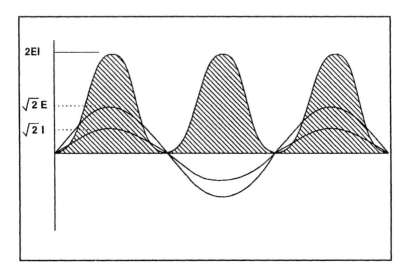

Figure 3-3. Power in a Resistive Circuit

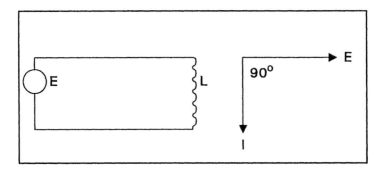

Figure 3-4. Phase Relationship in an Inductive Circuit

In the case of an inductance, the current I lags the voltage by 90° (see Figure 3-5). The plot of the sinusoidal waveforms shown below illustrates that the power P is positive for half a cycle and then negative for half a cycle. This means that power is oscillating back and forth between the generator (or capacitor) and the inductor and that average real power is zero. This type of power is defined as reactive power to distinguish it from the unidirectional active or real power previously discussed. The reactive power, given by the product EI, is measured in vars.

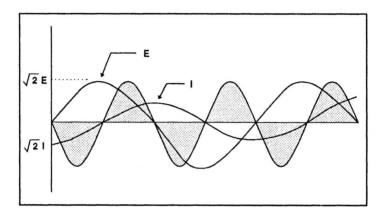

Figure 3-5. Power in a Totally Inductive Circuit

In a more realistic case, as shown in Figure 3-6, we will have both real and reactive power. Although the power is changing from a positive value to a negative value, we can see that the average power is positive, indicating a power absorbed by a resistive load. Power is defined as follows:

$$\text{Real Power} = P = E \cdot I \cos \theta$$

$$(\text{where } E \text{ and } I \text{ are given in RMS})$$

$$\text{Reactive Power} = Q = E \cdot I \sin \theta \qquad \text{3-1}$$

$$\text{Total Power} = \text{Complex Power} = S = P + jQ$$

$$\text{i.e., } S = VI \cos \theta + jVI \sin \theta$$

Relationship Between KVAR and Power Factor

The total power, given as KVA, delivered by the distribution system consists of both real (KW) and reactive power (KVAR). Reactive power, while it does no useful work, must still be supplied. The vector diagram commonly employed to express the relationship of real and reactive power is shown below in Figure 3-7.

67

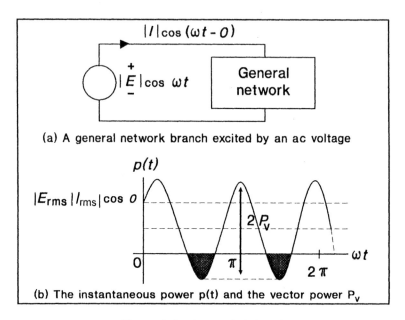

(a) A general network branch excited by an ac voltage

(b) The instantaneous power p(t) and the vector power P_v

Figure 3-6. General Network

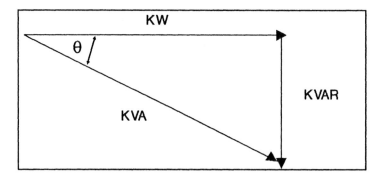

Figure 3-7. Real and Reactive Power Relationship

Where:

$$KVA^2 = KW^2 + KVAR^2$$

$$KW = KVA \cos \theta$$

3-2

$$KVAR = KVA \sin \theta$$

$$\cos \theta = \text{Power factor of the load}$$

Figure 3-8 shows examples of the change in kVAR for a 1000 kVA load as the power factor changes from .90 to .80.

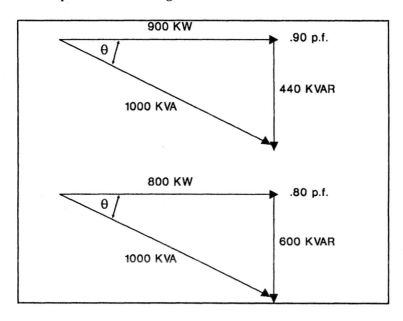

Figure 3-8. Examples of Effect of Power Factor

At 90% power factor, the kVAR supplied is approximately half of the kW. When the p.f. is reduced 10%, to .80, the kVAR becomes 75% of the kW component. If the power factor is reduced by another 10%, as shown in Figure 3-9 below, the kVAR increases to 100% of the kW component.

69

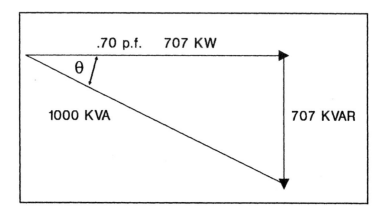

Figure 3-9. Example of Low Power Factor

An observation that can be made from these diagrams is that to change the power factor from .70 to .80 requires 107 kVARs while to change the power factor from .80 to .90 requires 160 kVARs. Likewise, it would take even another 440 kVARs to bring the load to unity power factor. The conclusion to be gained is that it takes more kVARs (a capacitor) to change a high power factor system than a low power factor system. This explains why it is almost always economical to add power factor correction on systems with power factors of .85 or even .90 but it becomes questionable whether a change above .98 or .99 is justified since the kVAR requirement is so much higher. It should also be noted that if the load is constant kW, then the effect of power factor correction is to reduce total kVA load. For example, if the 1000 kVA load originally had a power factor of .9, then the addition of capacitors to correct to unity power factor (440 kVAR) would result in a total kVA load of 900 kVA.

Examples

Exercise 1. As power factor on a 100 kVA load is lowered from .85 to .65 or approximately 24%, what percent change is there in kVAR?

Solution 1.

a.   p.f. = .85
b.   p.f. = .65

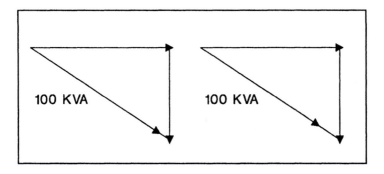

Figure 3-10. Example of Effect of Power Factor

p.f. = .85 = Ø = 32°
p.f. = .65 = Ø = 49.5°

a.   kW load = 100 × .85 = 85 kW
      kVAR load = 100 × .53 = 53 kVAR
b.   kW load = 100 × .65 kW = 65 kW
      kVAR load = 100 × .76 = 76 KVA

$$\% \text{ Decrease in p.f.} = \frac{.85 - .65}{.85} = 23.5\%$$

$$\% \text{ Increase in kVAR} = \frac{.76 - .53}{.53} = 43.4\%$$

∴ kVAR increased twice as fast as p.f. decreases

Exercise 2.

a.   How many kVARS of capacitors do I need to change the power factor from .8 to .85 on a 100 kVA load?

b.   How many kVARS of capacitors do I need to change the power factor from .9 to .95 on a 100 kVA load?

c.   What conclusions can you see?

Solution 2.

a.   100 kVA at .8 => kW = 100 × .8 = 80 kW
   kVAR = 100 × .6 = 60 kVAR
   100 kVA at .85 => kW = 100 × .85 = 85 kW
   kVAR = 100 × .5267 = 52.7 kVAR
   Needed kVAR = 60 - 52.7 = 7.3 kVAR
b.   100 kVA at .9 => kW = 100 × .9 = 90 kW
   kVAR = 100 × .436 = 43.6 kVAR
   100 kVA at .95 => kW = 100 × .95 = 95 kW
   kVAR = 100 × .312 = 31.2 kVAR
   Needed kVAR = 43.6 - 31.2 = 12.4
c.   •   Need more kVAR to change high power factor than low power factor.
   •   kW goes up the <u>same</u>.

Exercise 3. The combination of active and reactive power is sometimes referred to as complex power, defined as follows:

$$S = P + jQ$$

$$= VI \cos \theta + jVI \sin \theta$$

3-3

where "apparent power" is defined as the absolute value of S, i.e., S = VI. Complex power S, can more conveniently be defined as follows:

$$S = \tilde{V}\tilde{I}^* = \tilde{V}\left(\frac{\tilde{V}}{Z}\right)^* = \frac{\tilde{V}\tilde{V}^*}{Z^*} = \frac{|V|^2}{Z^*}$$

3-4

*implies the conjugate of a number, i.e., the same magnitude but an angle equal to an opposite.

Find the total complex power of the system and for each element in the system shown in Figure 3-11.

Suppose that, in Figure 3-11, the net impedance seen by the source E is

$$Z_{Total} = R_1 + jX_1 + \cfrac{1}{\cfrac{1}{-jX_2} + \cfrac{1}{jX_3} + \cfrac{1}{R_4}} \qquad \text{3-5}$$

$$= 110 + j\ 20 = 112\underline{/10.3^o}$$

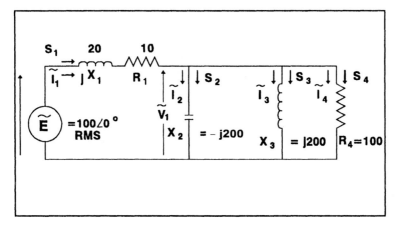

Figure 3-11. Power in a Typical Distribution Circuit

I, the total current, is then

$$\tilde{I} = \frac{\tilde{E}}{Z_{Total}} = 0.89\underline{/-10.3^o} \qquad \text{3-6}$$

S, the total complex power, is

$$S = \tilde{E}\tilde{I}^* = 100\underline{/0^o} \times 0.89\underline{/10.3^o} = 89\underline{/10.3^o} \qquad \text{3-7}$$

$$= 87.1 \text{ watts} + j15.8 \text{ vars}$$

$\cos \theta_1$, the power factor seen by the source, is

$$\cos \theta = \cos 10.3^o = 0.98 \qquad \text{3-8}$$

V, the voltage at the shunt load node, is

73

$$\tilde{V}_1 = \tilde{E} - \tilde{I}_1(R_1 + jX_1) = 89\underline{/-10.3^o} \qquad 3\text{-}9$$

S_2, the power consumed by X_2, is

$$S_2 = \tilde{V}_1 \tilde{I}_2^* = \tilde{V}_1 \left(\frac{\tilde{V}_1}{-jX_2}\right)^*$$

$$= \frac{\tilde{V}_1 \tilde{V}_1^*}{(-jX_2)^*} = \frac{|V_1|^2}{jX_2} = -j\frac{|V_1|^2}{X_2} \qquad 3\text{-}10$$

$$= -j\frac{89^2}{200} = -j39.8$$

i.e., the var consumption turning out negative means that vars are generated by the capacitive element.

S_3, the power consumed by X_3, is

$$S_3 = \frac{|V_1|^2}{(jX_3)^*} = \frac{|V_1|^2}{-jX_3} = j\frac{|V_1|^2}{X_3} = j\,39.8 \text{ vars} \qquad 3\text{-}11$$

i.e., the reactive element is consuming vars. Note that while this var "generation" and "consumption" convention is widely accepted, an alternative convention would result from defining S as

$$\tilde{E} * \tilde{I} = S_4, \text{ the resistive load is,}$$

$$S_4 = \frac{|V_1|^2}{R_4^*} = \frac{|V_1|^2}{R_4} = 79.2 \text{ watts} \qquad 3\text{-}12$$

The concept of reactive losses in series elements is important too. In X_1,

$$S = P + jQ = |I_1|^2 (jX_1) = 0 + j15.8 \text{ vars} \qquad 3\text{-}13$$

a purely reactive loss.

In R_1,

$$S = P + jQ = |I|^2 (R_1) = 7.92 + j0 \text{ watts} \qquad 3\text{-}14$$

a purely resistive loss.

The total losses for the circuit are then

	P	Q
R_1	7.9	0
X_1	0	15.8
X_2	0	-39.8
X_3	0	+39.8
R_4	79.2	0
TOTAL	87.1	15.8

Note that the total agrees with the original solution for S, total complex power supplied to the circuit.

Exercise 4. What happens if we disconnect capacitor X_2 from the circuit shown in Exercise 3?

Now,

$$Z_{total} = R_1 + jX_1 + \cfrac{1}{\cfrac{1}{jX_3} + \cfrac{1}{R_4}}$$

$$= 10 + j20 + \cfrac{1}{\cfrac{1}{j200} + \cfrac{1}{100}} \qquad 3\text{-}15$$

$$= 10 + j20 + \frac{J200}{1 + j2}$$

$$= 10 + j20 + 80 + j40$$

$$= 90 + j60 = 108\underline{/33.7^\circ}$$

I, the total current, is then

$$\tilde{I} = \frac{\tilde{E}}{Z \text{ Total}} = \frac{100/0}{108/33.7} = .926\underline{/-33.7} \qquad 3\text{-}16$$

S, the total complex power is,

$$S = \tilde{E} \ I^* = 100\underline{/0^o} = .926\underline{/33.7}$$

$$S = 92.6 \underline{/33.7} \qquad 3\text{-}17$$

$$P = 77.0 \text{ Watts}$$

$$Q = 51.4 \text{ Vars}$$

and V_1, the voltage across the load, is

$$\tilde{V}_1 = \tilde{E} - I_1 \ (R_1 + jX_1)$$

$$= 100\underline{/0^o} - .926\underline{/-33.7} \ (10 + j20) \qquad 3\text{-}18$$

$$= 100\underline{/0^o} - 20.7\underline{/29.7} = 100 - 17.93 - j10.35$$

$$= 82 - j10.35 = 82.7 \underline{/-7.15}$$

Some of the interesting observations which can be made and give some insight into the value of capacitor banks are as follows:

a. Current goes up after the capacitors are removed.
b. Watts or real power consumption goes down.
c. VAR consumption goes up.
d. Voltage to the load goes down.

VOLTAGE RISE AND VOLTAGE DROP

Early distribution lines were generally voltage limited. This means that because the lines were usually long and the primary voltage distribution low, the voltage drop became so great as to curtail further load additions to the line. Most of these lines relied heavily on the use of voltage regulators to provide customer voltage within the prescribed limits.

Today, most distribution lines are thermally limited, i.e., the maximum current the line can carry without burning down is reached before voltage drop became a problem. This is possible because modern distribution systems generally have shorter feeders and higher voltages (15, 25, and 35 kV). Voltage drop on these lines certainly still exists, but with most utilities extensive use of capacitors limits line loss. Many utilities have found that the judicious sizing and location of these capacitors will provide not only loss reduction but also all the voltage control they need without the use of feeder regulators.

Voltage Drop Defined

When the electrical characteristics (R&X) of the line have been determined and the power factor of the load is known it is very straightforward to calculate voltage drop. Let's take a look at the system shown in Figure 3-12.

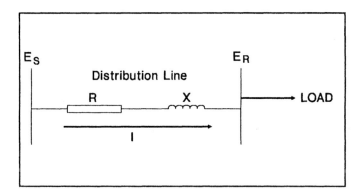

Figure 3-12. Voltage Drop in a Distribution Line

If we assume that the load is reactive, such as a motor, we would expect that current would lag the voltage.

The voltage at the sending end, E_S, can be calculated as follows:

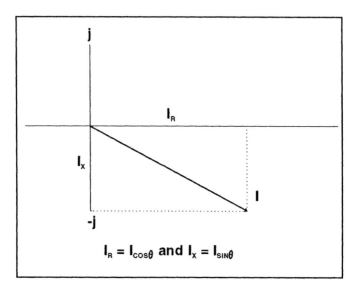

Figure 3-13. Phase Relationship for an Inductive Load

$$E_S = E_R + \text{LINE DROP}$$
$$= E_R + \tilde{I}(Z)$$
$$= E_R + \tilde{I}(R + jX)$$
$$= E_R + (I_R - jI_X)(R + jX)$$
$$= E_R + I_R \cdot R - jI_X \cdot R + jI_R \cdot X + I_X \cdot X$$
$$= E_R + I_R \cdot R + I_X \cdot X - jI_X \cdot R + jI_R \cdot X$$

3-19

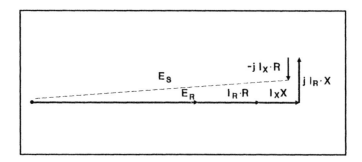

Figure 3-14. Vector Diagram for Voltage Drop

Actual voltage drop is equal to $E_s - E_R$. If we project E_S to the E_R axis, we see that the true voltage drop is almost equal to $I_R \cdot R + I_x \cdot X$ and the out-of-phase voltage drop components (i.e., $-jI_x R$ and $jI_R \cdot X$) have virtually no effect on the total. For this reason, the following equation is valid for most applications:

$$\text{Voltage drop} = I(R \cos \theta + X \sin \theta) \qquad 3\text{-}20$$

This formula gives the voltage drop on one conductor, line-to-neutral. The three-phase line-to-line drop is $\sqrt{3}$ times the above value, and the single-phase drop is twice the above value.

The addition of a capacitor to an inductive system decreases voltage drop since the in-phase drop produced across the line by the capacitor is in the opposite direction to the voltage drop produced by load across the resistance and reactance.

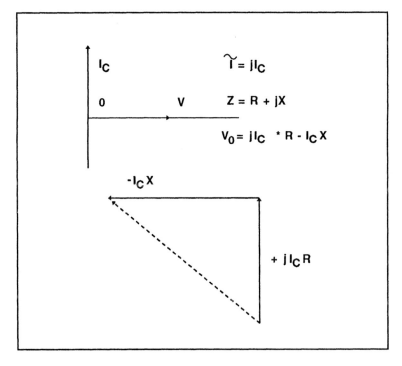

Figure 3-15. Vector Diagram for Capacitor Effect

Voltage Rise

Lagging, or reactive current, usually causes a greater voltage drop than an increase in active current. This is primarily due to the fact that the reactance of a line is generally greater than the resistance.

Figure 3-16 illustrates graphically the relative effects of the active and the reactive components of the current on voltage. From this figure, it is seen that over a five mile length of typical No. 4/0, 4160 voltage distribution circuit, 200 active amperes cause a voltage drop along the circuit of 11.5% of the total voltage at the sending end. On the other hand, 200 reactive amperes lagging will produce a drop of 28.6% of the total voltage at the sending end. Therefore, for this typical circuit, the destructive effect of lagging reactive amperes on useful voltage is 2.5 times that of the active current.

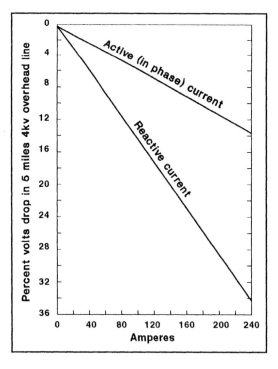

Figure 3-16. Comparative Voltage Drops Caused by Active and Reactive Currents

Shunt capacitors, as explained earlier, are utilized to counteract this voltage drop by producing a "voltage rise" across the line. Shunt capacitors applied to a feeder give a voltage rise which is <u>fixed</u> (for all practical purposes), and is <u>independent</u> of the magnitude or power factor of the load. The formula for this rise is given as:

$$\% \text{ Voltage Rise } = \frac{\text{CKVA} \times X \times d}{10 \times \text{kV}^2} \qquad \text{3-21}$$

where: CKVA =   the 3∅ rating of the capacitor bank
             X =   the line reactance per mile
             d =   distance from substation
          kV =   line-to-line voltage

If we have a 34.5 kV system with a 900 kVAR bank of capacitors, for example, we would calculate the rise as follows:

$$\% \text{ Voltage Rise } = \frac{900 \times .5 \times 10}{10 \times (34.5)^2} = .38\% \qquad \text{3-22}$$

Switched vs. Fixed Capacitors

If we have a distribution system with no load connected to it (and we neglect magnetizing and charging currents), we will have no current and consequently no voltage drop. If we modify this system, as shown in Figure 3-17, by placing a shunt capacitor on it, we have now created a distribution system with a leading power factor and a voltage rise.

Using the formula for voltage drop, we can verify that the voltage at the capacitor bank is actually higher than the source voltage E_S.

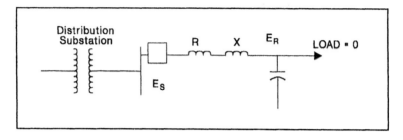

Figure 3-17. Effect of Leading Power Factor

$$E_R = E_S - \text{Voltage Drop}$$

$$\text{V.D.} = I(R \cos \theta + X \sin \theta)$$

$$\theta = -90^\circ \Rightarrow \cos(-90^\circ) = 0 \qquad\qquad 3\text{-}23$$

$$\sin(-90^\circ) = -1$$

$$E_R = E_S - I(-X) = E_S + IX$$

If the IX calculated is large enough, we might have a voltage at the end of the system which is "out of standards" and too high at the customer service entrance. This is one of the reasons that the distribution system should not allow the power factor to exceed unity.

On a typical distribution system, the load varies as a function of the time of day. Since the load at certain hours is very low, for example, in the early hours of the morning and very high at other times (e.g., 9 AM and 5 PM), it is obvious that voltage drop is different as are the VAR requirements.

Figure 3-18 shows the kW and MVAR requirements as a function of time for one particular utility.

The question which should be asked is: "If the VAR requirements are always changing, how many kVARs should I put on my system to maximize my investment and not create a voltage problem?" The optimum solution would be to have a VAR supply system that would track the VAR demand exactly. Although this is possible to do, it is much too expensive. What many utilities have done in an effort to compensate for the changing VAR requirements is to utilize both fixed and switched capacitor banks. For the load cycle shown in the previous figure, we might put in a fixed bank of capacitor of 300 kVAR to compensate for light load and a switched bank of another 300 or 400 kVAR to compensate for heavy loading conditions. It should be noted that even with this scheme, it is possible to have a leading power factor for short periods during the day.

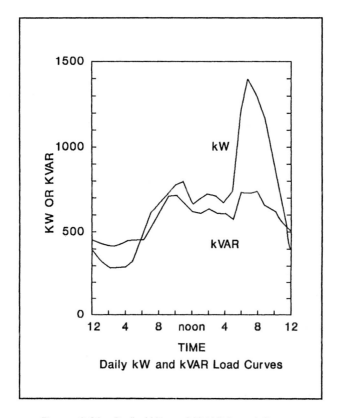

Figure 3-18. Daily kW and KVAR Load Curves

This is generally acceptable because the net voltage rise is usually not a problem during the peak loading hours.

Types of Capacitor Control

Once the system kVAR loading requirements in respect to time are known, it can be determined what amount of capacitive kVARs should be installed as fixed banks to compensate for light load periods and what amount of kVAR loading can be balanced by switching banks to compensate for peak load periods. What is attempted is to ideally obtain close to unity power factor during all periods of time by adding capacitive kVAR to the system when required and to remove the capacitive kVARs from the system when not required so as to prevent leading kVARs to be "seen" by the generator, which may cause overexcitation and possible instability. The more modern generators are much less susceptible to this problem.

Reference to Figure 3-19 below shows adding capacitors to a feeder, or system, can reduce the lagging load kVARs which the supply must deliver. In this case, source kVARs are never leading and during peak load periods the source is delivering approximately one-half of the kVARs it would have had to supply if no capacitors had been added. It should be noted that the off-setting of each control is lower than the reduction in kVARs obtained when the bank it is controlling is switched on. This is necessary in order to prevent "hunting" or "pumping". This is characteristic of controls that sense electrical conditions, such as voltage, current or VAR. Controls operated by time or temperature are not subjected to this problem.

Where the capacitors are to be located on the system is determined by maximum loss reduction. How the capacitors are to be controlled is, of course, dependent on the cause and nature of the load.

Switched capacitors installed on distribution feeders and laterals usually have less than 1800 kVAR ratings, 300 to 1200 kVAR being the most popular. One purpose is to provide voltage regulation. Higher levels of voltage can be maintained with resultant higher revenues. Lower voltages can be maintained if fuel conservation is of prime concern due to lower energy consumption by lighting loads. In all cases, most public utility commissions limit the level of voltage supplied to customers to ±5%, based on 120 volts as normal (114-126 volts). Most utilities try to limit the variation in voltage to less than 8%, some to less than 6%, while still remaining within the limits set by the PUC.

The change in voltage due to switching capacitor banks on and off the lines produces a "ripple effect" both upstream and downstream from where the bank is located. Therefore, all controls in the area of a bank

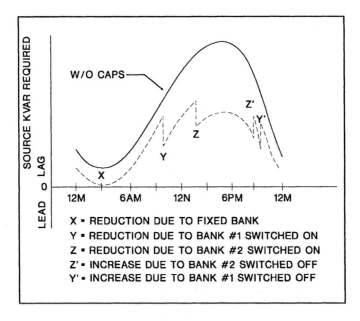

Figure 3-19. Effect of Capacitor KVAR's on Source Requirement

being switched will see that on and off settings are adjusted to take this ripple effect into consideration.

When it becomes difficult or impossible to set the operating points on a voltage control because of the voltage changes due to other banks being switched, other means may have to be used such as using current sensing controls or time switches, etc. These will also cause the voltage variation to be less. Reference to the figure below shows the effect of switching capacitors on a feeder, by whatever means, and the resultant decrease in voltage variation. Figure 3-20 portrays the voltage profile at one point on the daily load curve, say, at peak load, of a fictitious feeder.

Maximum permissible voltage rise due to switching a bank on varies from 2% to 5%, depending on the type of circuit and number of switching operations per day. The total rise at the instant the bank is switched on is dependent on the circuit impedance all the way back to the source and the decrease in line current due to the reactive current supplied by the bank.

85

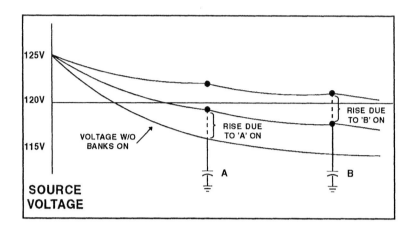

Figure 3-20. Feeder Voltage Profile at One Point on Daily Load Curve

There are several popular methods of controlling switched banks of capacitors. Each method has advantages and disadvantages and many utilities use combinations of these basic methods to optimize capacitor performance. The standard methods are:

1. Voltage - this method is relatively inexpensive and works well where voltage varies with load. On short feeders where voltage drop is not great or on feeders with regulators, this method may be difficult to coordinate. It is many times used to supplement other methods for emergency conditions.

2. Current - this method responds to loading well and works well for systems having induction voltage regulators. It does, however, require a current sensing device (expensive) and cannot differentiate between low power factor summer loads and high power factor winter loads.

3. Kilovar - this method is most effective for minimizing IX losses and can also differentiate between summer and winter peak loading conditions. It is, however, very expensive and provides only fair voltage control when used out on the system.

4. Time - this method is simple and inexpensive. It does not sense abnormal loads and can often get out of sync due to extended power outages, holidays, etc.

5. Temperature - the advantage of this method is that it's simple and low cost. On systems where VAR load varies with temperature (air conditioning) it works well. It does not, however, sense holidays and often requires additional controls to maintain acceptable voltage levels.

Exercise 5. If voltage is defined as $V_D = I$ (R cos + X sin), what happens to the voltage when a 5 mile piece of conductor with a loop impedance of .053 + J.095 ohms/1000 ft. sees the load changed from 200 amperes real to 200 amperes lag?

Solution 5.

$$R = .053 \times 5.28 \times 5 = 1.3992$$

$$X = .095 \times 5.28 \times 5 = 2.508$$

3-24

a. 200 Amps Active

$$V_D = 200 \ (1.3992 \times .1) = 280 \ \text{Volts}$$

3-25

b. 200 Amps Reactive

$$V_D = 200 \ (2.508) = 502 \ \text{Volts}$$

3-26

Conclusion. The detrimental effect of lagging amperes is greater than real amperes. On a 4160 system, this is equivalent to:

$$\frac{280}{4160/\sqrt{3}} \times 100\% = 11.66\%$$

3-27

$$\frac{502}{2400} \times 100\% = 20.9\%$$

LOSSES

Line Loss Defined

Over the years since the oil embargo, the subject of losses has received a great deal of attention. When the cost of electrical energy was low, utilities did not concern themselves as much with the line loss (and transformers) of their system. The result of this was that many distribution feeders were very long, relatively low voltage (5 kV) and were usually voltage drop limited. Today, with higher density loadings, it is necessary to have higher voltages (15, 25, 35 kV) and shorter lines resulting in lines that are now thermally restricted. Since losses are a function of the square of the current and are very costly, the application of capacitors to reduce these losses has become imperative.

The peak of kW losses L are caused by the current I squared, and flowing through the resistance R, of the line.

$$L = kW \; loss = \frac{I^2 R}{1000} \qquad\qquad 3\text{-}28$$

where $I = I \cos \theta + jI \sin \theta$ or $I = \sqrt{(I \cos \theta)^2 + (I \sin \theta)^2}$

$$L = \frac{R}{1000} [(I \cos \theta)^2 + (I \sin \theta)^2] \qquad\qquad 3\text{-}29$$

if we define L'as the real (in phase component) $= R(I \cos \theta)^2$
and L" as the reactive (out of phase component) $= R(I \sin \theta)^2$

We can now determine the % load losses due to kVARS = L"/L.

$$\text{or} \quad \frac{\dfrac{R}{1000} (I \sin \theta)^2}{\dfrac{R}{1000} [(I \cos \theta)^2 + (I \sin \theta)^2]} \qquad\qquad 3\text{-}30$$

$$\text{dividing by } \frac{RI^2}{1000} \qquad\qquad 3\text{-}31$$

$$\frac{L''}{L'} = \frac{(\sin\theta)^2}{(\cos\theta)^2 + (\sin\theta)^2} = \sin^2\theta \qquad\qquad 3\text{-}32$$

A plot of this relationship is shown below:

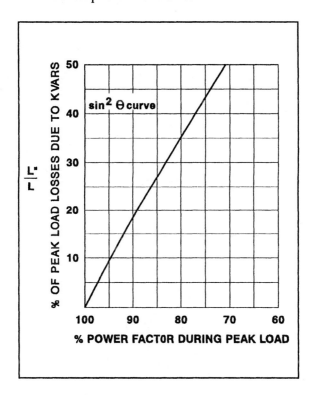

Figure 3-21. *Losses Due to KVARS*

As mentioned earlier, the most important formula to determine loses is L = $I^2R/1000$ (in kW). Consequently, to reduce losses, you must reduce the R (loop resistance $\approx$ 2 times conductor resistance) or reduce the absolute value of I. Since I is equal to I cos θ + I sin θ we can see that the reduction of the reactive component of current has a direct effect on "real" power loss. Said another way, the higher the power factor the lower the magnitude I.

For example, if we had a 10 mile line with a loop impedance of .3 + j.7 ohms/mile with a load located at the end of the line of 400 amps, the losses of the line would be calculated as follows:

$$L = \frac{I^2R}{1000} = \frac{(400)^2 \times 10(.3)}{1000} = 480 \text{ kW} \qquad \text{3-33}$$

If the load were evenly distributed, similar to the diagram shown in Figure 3-22, we would have much less loss because the current I decreases with distance. The equivalent losses are when the total load is placed 1/3 of the distance from the substation. Total losses are then calculated as:

$$\frac{(400)^2 \times \frac{10}{3}\,(.3)}{1000} = 160 \text{ kW} \qquad \text{3-34}$$

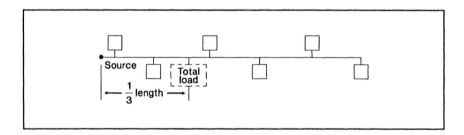

Figure 3-22. Uniformly Distributed Load

Another typical feeder configuration is shown below. This condition is normally found when the line coming out of the substation is an express portion of line or cable feeding a load some distance away. In this instance the total loss is calculated as follows: Assume

$$L_1 = 5 \text{ miles}$$

and

$$L_2 = 5 \text{ miles}$$

3-35

$$\text{Total Loss L} = \frac{(400)^2 \ (.3) \ (5 + \frac{5}{3})}{1000}$$

$$= \frac{(400)^2 \ (.3) \ (6.66)}{1000}$$

$$= 320 \text{ kW}$$

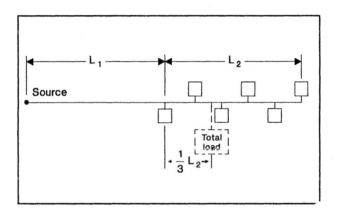

Figure 3-23. Uniformly Distributed Load Over Part of Line

In the previous example we did not assume a power factor because only the magnitude of the I determines the losses. However, if we made the assumption that our 400 amperes had a power factor of .85 and we could use a capacitor to compensate for the power factor we could see the value of this investment. For an uncompensated 400 amperes load at the end of our line the losses would be

$$L = \frac{400^2 \times (.3) \times 10}{1000} = 480 \text{ kW} \qquad \text{3-36}$$

If the load had a .85 p.f., the current is now comprised of both real and reactive components.

$$\text{i.e., } I_{REAL} = 400 \cos \theta = 400 \times (.84) = 340 \text{ amperes}$$
$$\text{3-37}$$
$$I_{REACTIVE} = 400 \sin \theta = 400 \times (.52) = 208 \text{ amperes}$$

Since our capacitors can eliminate all of the reactive component (i.e., the 208 amps) we are left only with the real component or the 340 amperes. The total loss would then be

$$\frac{(340)^2 \ (.3) \times 10}{1000} = 347 \text{ kW} \qquad \text{3-38}$$

Consequently, the reduction in losses due to the correction of power factor is 480-347 or 133 kilowatts.

Optimal Capacitor Placement

The placement of the capacitor on the distribution system is very, very important. For example, a capacitor placed at the substation has no effect on line losses since the reactive kVA must still be sent all the way from the substation to the load as shown below:

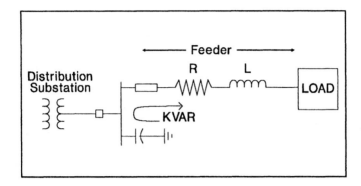

Figure 3-24. Substation Capacitors

For the condition shown, the optimum placement of the capacitor bank is obviously at the load. But where should the capacitors be placed if the load is distributed? Suppose we assume a distribution line which has only resistance R and an evenly distributed totally inductive load of I modeled at 4 locations (see Figure 3-25). We can easily calculate the I^2R losses of this circuit as follows:

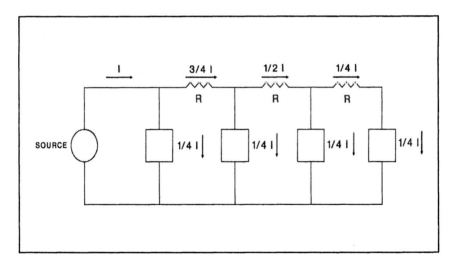

Figure 3-25. Distributed Load (I)

$$L = (3/4\ I)^2R + (1/2\ I)^2R + (1/4\ I)^2R$$

$$= 9/16\ I^2R + 4/16\ I^2R + 1/16\ I^2R \qquad 3\text{-}39$$

$$= \frac{7}{8}\ I^2R \approx I^2R$$

If we look at the effectiveness of capacitor compensation, we see that it is very important where we place the bank of capacitors. For example, let's assume that we place a capacitor bank 3/4 of the total load (I) at the end of the feeder:

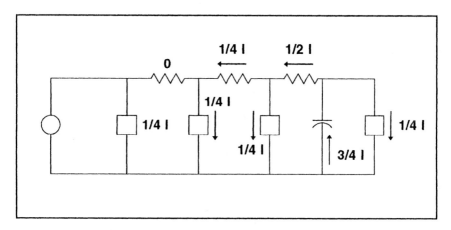

Figure 3-26. Effect of Capacitor at End of Line

The total loss =

$$O^2R + (1/4\ I)^2R + (1/2\ I)^2R = (1/16 + 1/4)I^2R \qquad 3\text{-}40$$

$$= 5/16\ I^2R$$

which is over a 50% reduction in losses.

If we move the capacitor bank, one section closer to the source we have the following situation:

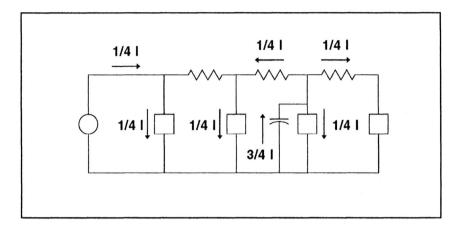

Figure 3-27. Optimum Location

The losses are:

$$0 + (1/4 \text{ I})^2 R + (1/4 \text{ I})^2 R = 1/8 \text{ I}^2 R \qquad 3\text{-}41$$

or less than 20% of the original system losses for the <u>same amount</u> of capacitors. This example points out why it is so important for a utility to consider where the capacitors are placed. It also shows, when taken to its limit (for a distributed load), the validity of the "2/3's rule", i.e., to minimize losses we should place 2/3 of the needed kVAR 2/3 of the way down the feeder (see Figure 3-28). Any deviance from this will only increase losses. Try it!

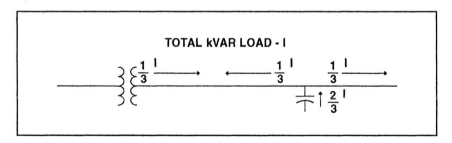

Figure 3-28

Load Factor and Loss Factor

Load factor is defined as the ratio of Average KW Demand divided by Peak KW Demand (Average/Peak). It is understandable that there should be a relationship between system load factor which most utilities know and line losses, which are much harder to calculate.

Line losses, which are the sum of the I^2R, or resistance losses, can be determined when the currents at peak load are known. Since the current in an electrical system varies with respect to time, there is no precise method to calculate losses over a period of time. However, it is done by multiplying peak loss by loss factor. Loss factor is usually defined as the ratio of the average power loss over a designated period of time, to the maximum loss occurring in that period. Definitions of loss factor and load factor are quite similar and care should be taken that the matter is not used in place of loss factor when considering system losses. There is a relationship between the two factors and the empirical relation at the distribution transformer is given by

$$LsF = 0.15 \ LdF + 0.85 \ LdF^2 \qquad\qquad 3\text{-}42$$

Suppose we calculate our peak kW line losses as 180 kW. If the system had a load factor of .7, then the average kW losses could be calculated as follows:

$$180 \times [.15(.7) + .85(.7)^2] = 180(.5215) = 94 \ kW \qquad 3\text{-}43$$

Exercise 6. An 8 mile, 34.5 kV utility line with an impedance of .2 + j.6 ohm/mile serves an industrial customer with a peak load of 20 MVA at .8 PF and a .6 load factor. The customer has just purchased a 3 MVAR bank of capacitors and insists he be given full credit on his annual bill for the returned line losses on the utility system. How much could he legitimately be entitled to at an energy cost of $.08 per kW hr?

Solution 6.

$$\frac{\dfrac{20 \ MVA}{3 \times 34.5}}{\sqrt{3}} = 334 \ amps \ peak \qquad 3\text{-}44$$

$$\text{Loss Factor} = .15(.6) + .85(.6)^2$$

$$\text{Avg. Loss} = I^2R \times \text{Loss Factor} \qquad \text{3-45}$$

$$= (334)^2 \times (8 \times .2) \times .396$$

$$= 70.682 \text{ kW per phase}$$

Total Annual Losses =

$$3 \times 8760 \times 70.682 = 1,857,523 \text{ kwhr} \qquad \text{3-46}$$

$$\text{Cost/Year} = \$148,602/\text{yr}$$

Losses with Capacitor Bank =

$$20 \text{ MVA at .8 PF} = 16 \text{ MW}$$

$$12 \text{ MVAR}$$

$$\text{with Capacitor Load} = 16 \text{ MW and 9 MVAR}$$

$$= \text{New Load is 18.36 MVA}$$

$$\cos = \frac{16}{18.3} = \text{p.f.} = .87 \qquad \text{3-47}$$

$$I = \frac{18.36}{\sqrt{3} \times 34.5} = 307.25 \text{ amps}$$

$$\text{Avg. Loss} = (307.25)^2 (8 \times .2) \times .396$$

$$= 59.81 \text{ kW per phase}$$

Total Annual Cost of Losses =

$$3 \times 8760 \times 59.81 \times .08 = \$125,745 \qquad \text{3-48}$$

Cost savings:

$$
\begin{array}{r}
148{,}602 \\
-\ 125{,}345 \\
\hline
\$\ \ \ 22{,}857
\end{array}
$$

∴ *You could reduce his annual bill by $22,857 per year*

Exercise 7. A distribution line 10 miles long has an impedance of .3 + j.7 ohms/mile. Other characteristics of this line are:

Load factor = 6
Load is distributed
Cost of energy is $.05/kWHR
Peak load = 400 amperes
Power factor = .8

How much can a utility save in energy costs by correcting the p.f. from .8 to unity?

Solution 7.

a. Calculate loss factor:

$$= .15 \text{ Load Factor} + .85 \text{ Load Factor}^2 \qquad \text{3-49}$$

$$= .15\ (.6) + .85\ (.36) = .4$$

b. Calculate real and reactive amps:

$$\text{Real amps} = 400 \times .8 = 320 \qquad \text{3-50}$$

$$\text{Reactive amps} = 400 \times .6 = 240$$

c. Calculate average losses for present system:

$$= I^2R = 400^2 \times .3 \times .4 \times 10$$

$$= 192{,}000 \text{ watts}$$

3-51

$$= 192 \text{ kW} *$$

* 3 phases × 1/3 factor for evenly distributed load = 1

d. Calculate I at unity power factor:

$$\text{I at P.F.} = 1.0 = 400 \times .8$$

3-52

$$= 320 \text{ amperes real and 0 reactive}$$

e. Calculate average new loss:

$$I^2R = 320^2 \times .3 \times 10 \times .4 = 123 \text{ kW}$$

3-53

f. Calculate cost of losses:

$$= (192 - 123) \times 8760 \times .05 = \$30{,}222$$

3-54

QUESTIONS

1. Why does imaginary power do no work?

2. A load having 2 megawatts at a power factor of .87 has how many kWs and kVARs?

3. Why does it cost more to increase power factor by capacitors when the power factor is higher?

4. Voltage rise due to a capacitor bank is dependent on what two things?

5. Switched capacitors are used for what two reasons? Explain what happens when power factor is too high.

6. Reactive power can cause more voltage drop than real power (true or false)? Explain.

7. What is the reason not to use a switched capacitor bank controlled by voltage?

8. Why does reactive load cause losses?

9. Optimum capacitor placement for an evenly distributed load occurs when half the VARs from the capacitor flow toward the source and half flow away from the source. Is this consistent with the 2/3's rule?

10. A 900 kVAR bank of capacitors could rarely cause excessive voltage rise on a 3 mile long feeder. Explain.

4

DISTRIBUTION OVERCURRENT PROTECTION

INTRODUCTION

The overcurrent protection of the distribution system is considerably different from any other part of the utility system. Unlike the transmission and subtransmission system, the distribution system is usually radial in design. Also, where the protection of the transmission system is usually performed by breakers with various types of relays, the distribution system utilizes breakers, circuit switchers, load break disconnects, fuses, overcurrent relays, reclosers and sectionalizers (see Figure 4-1).

The fact that virtually all distribution systems are radial is what allows the protection system to work properly. A radial system is coordinated on the premise that the fault current decreases as distance from the substation increases. A profile of fault current vs. distance is shown in Figure 4-2 and illustrates this relationship. As can be seen, a line-to-ground fault at the terminals of the substation would create fault currents on the order of 6500 amps. Just one mile away, this fault current is reduced to less than 3000 amperes. A fault 10 miles from the substation will draw less current than full load conditions indicating that protection of faults far out on the feeder cannot be performed at the substation.

CHARACTERISTICS OF DEVICES

Fuses

Fuses are the most basic and cost effective type of overcurrent device presently being used by the utility industry. They are also one of the most reliable devices in that they can provide their function for over 20 years with essentially no maintenance.

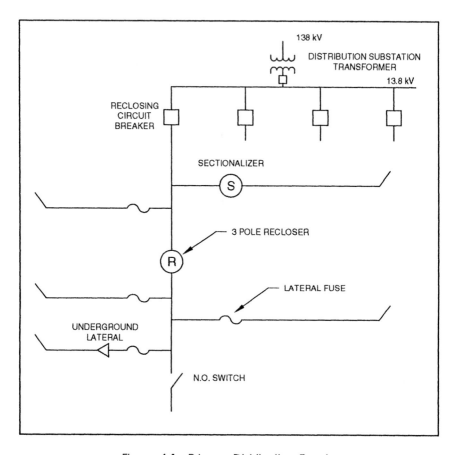

Figure 4-1. Primary Distribution Feeder

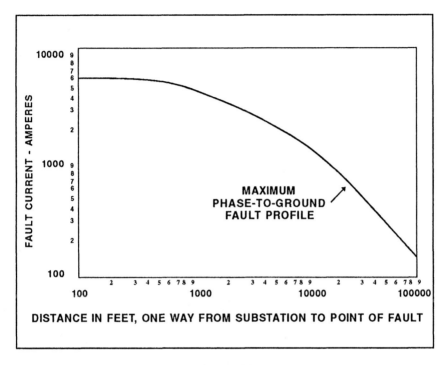

Figure 4-2

All overcurrent protecting devices carry the more commonly used name of "current interrupter". Current interrupters do not actually interrupt current. What they actually do is setup a high dielectric environment that prevents the arc from re-establishing when the current passes through a current zero.

Fuses use the following mechanism to interrupt the flow of current:

1. Sensing . . . . . heat and melt
2. Arc initiation . . . . . separation
3. Arc manipulation . . . . . stretching, cooling, deionizing, pressurizing
4. Current interruption . . . . . current zero.

Although a fuse is deceptively simple in appearance, its function is complex. For a fuse to function properly, it must:

103

- Sense the conditions it is trying to protect
- Interrupt the fault quickly
- Coordinate with other protective devices.

Expulsion Fuses. A distribution expulsion fuse link consists of three basic parts which are:

- Button
- Fusible element
- Leader

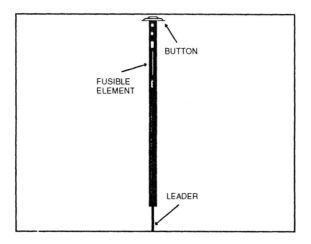

Figure 4-3. Fuse Link

Length and diameter of the element are the main determinants of the fuse characteristics. The longer the fuse, the <u>faster</u> it will operate for low level short circuits. During high levels of short circuit, the temperature rise is very fast and heat cannot be conductored away from the center of the fuses element. Consequently, melting time at the higher current range is not dependent so much on the length of the element but rather by the diameter.

The expulsion fuse, as the name implies, expulses gases during its operation. Figure 4-4 shows a diagram of a typical expulsion fuse cutout. The fuse itself is housed inside the fuse cartridge.

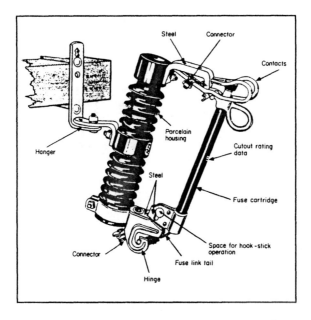

Figure 4-4. A Distribution Expulsion Circuit

The typical expulsion fuse will utilize a relatively short fusible element section to sense the overcurrent and start the arcing required for interruption. Attached to this short fusible element will be a larger type of conductor, commonly called a fuse leader or fuse link tail, which then connects to the rest of the fuse hardware as required. During a fault, the fuse element will melt causing an arc inside the fuse cartridge. When the arc is produced it will rapidly create gases from special materials (usually fiber) located in close proximity to the fuse element. The primary function of the gases that are generated by the fiber is to deionize and remove arc generated ionized gases and allow a rapid buildup of dielectric strength that can withstand the transient recovery voltage and steady state power system voltage. Of course, consideration must be given to the fact that these gases will be produced and will be exhausting from the fuse assembly. Some types, such as the boric acid fuses, can utilize condensers to prevent much exhaust beyond the fuse assembly.

One of the main advantages with typical expulsion fuses is that they can be reloaded with a relatively economical fuse link. Furthermore, there are a wide variety of fuse link types and sizes that could be used in the

105

same fuse holder. This allows common usage of ⟨
number of applications and also permits a wide latitu
possibilities.

EEI-NEMA standards divide expulsion fuse links in⟨
fast and slow designated by K and T, respectively. K and 1
same rating have identical 300 second point or 600 second poin⟨
shown in Figure 4-5, they have different T/C curves, the T link
slower at the high-current end than the same size K link.

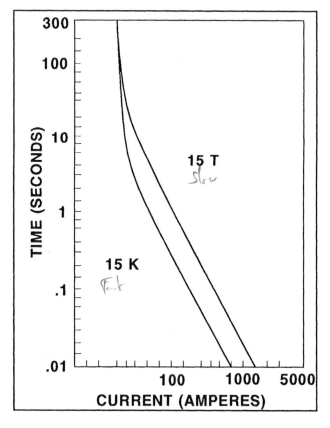

Figure 4-5. Minimum Melting Time Curves for
Fast and Slow Links of the Same Rating

The distinction between the two types is the speed ratio, which is the ratio between melting currents at 0.1 second and 300 seconds for links rated through 100 amperes, and at 0.1 second and 600 seconds for links rated over 100 amperes. For example, a T link rated at 6 amperes has a 0.1 second melting current of 130 amperes and a 300 second melting current of 12 amperes, resulting in a speed ratio of 10.8. Slow links have speed ratios between 10.0 and 13.0. Fast links have speed ratios between 6.0 and 8.1.

To meet special requirements such as primary fusing of small-sized transformers, links rated below 10 amperes have been developed. High-surge links rated 1, 2, 3, 5, and 8 amperes are in this category. They are specifically designed to provide overload protection and avoid unnecessary operation during short-time transient current surges associated with motor starting and lightning.

Current Limiting Fuses. The two key components in the construction of a current limiting fuse are the fusible element, which is normally of pure silver and quartz sand. Thus, the current limiting fuse is sometimes called a silver-sand fuse.

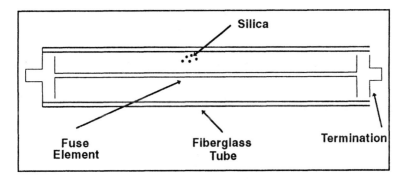

Figure 4-6. Current Limiting Fuse

Silver is used in the current limiting fuse because it has the best balance of properties to produce the highest ampere rating for a minimum physical size and also has the most desirable melting characteristic. The silver element will not become damaged under overcurrent conditions until very close to its high melting temperature of 960°C. When it reaches this melting temperature, only a small amount of additional heat is required to completely melt and vaporize the metal. This characteristic

107

is desirable to minimize the total energy released in the fuse during an interruption. In addition, silver has excellent current carrying ability and immunity to oxidation.

The silica sand functions to absorb the heat given off by the fusible elements without becoming electrically conductive. Though this heat absorbing function is basically simple, the presence of the sand is critical to the correct operation of the fuse. Since the temperature of the arc approaches that of the sun, few materials can perform this heat absorbing function successfully. Even seemingly incidental considerations such as the size of the sand granules and the method of packing can have significant influence on the operating characteristics of the fuse.

More specifically, a current limiting fuse consists of one or more silver wire or ribbon elements (one manufacturer uses cadmium) suspended in an envelope filled with sand. These silver elements are spirally wound on a temperature-resistant, non-tracking form or core. By spirally winding the silver elements, the length of the fuse can be kept to a minimum. The core and element assembly is mounted in a tube of temperature-resistant material and then filled with the high-purity silica sand and sealed. The degree of sealing may vary with the application. For example, fuses designed for direct oil submersion must be carefully sealed to prevent the ingress of oil at temperatures as high as 150°C. Fuses for outdoor application must be capable of withstanding the outdoor environments and be sealed against possible entry of moisture over the service life, which is generally 20 years or longer.

Both wires and ribbon-type fusible elements or combinations of the two are used in the construction of current-limiting fuses. When silver ribbons are used, each ribbon has small holes punched along its length forming a series of notches. These notches control heat distribution and help shape the time-current characteristics. In normal operation, the I^2R heat generated flows into the relatively large and cool body of the element, then into the surrounding sand. When a fault occurs, heat is generated so rapidly that almost none can escape. Instead, it remains in the elements and melts the silver elements.

On high-fault currents, the element melts instantaneously along its entire length to interrupt the current. Upon interruption, the heat of the arc is transferred to the sand, transforming it into a crystalline structure known as a fulgurate. This transformation results in the sudden insertion of additional impedance and the subsequent development of an arc voltage. When the arc voltage exceeds the system (driving) voltage, the arc is extinguished and current-limiting action is accomplished. In the figure below, note the magnitude of the developed arc voltage. Since it exceeds the system voltage, its effect on surge arresters must be analyzed. The higher the generated voltage, the greater the current-limiting action.

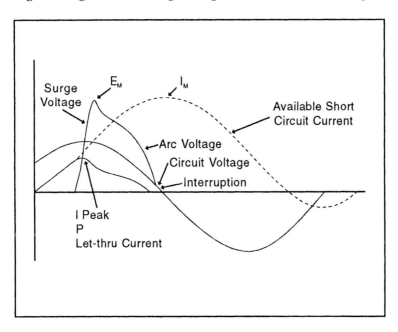

Figure 4-7. Voltage Current Relationships During Operation of a Current-Limiting Fuse in a Zero Power Factor Circuit

Current-limiting fuses are available in two types: general purpose and backup. Both employ a silver fuse element wound on a supporting core.

The general purpose fuse is defined by ANSI C37.40 - 1969 as follows:

> "A fuse capable of interrupting all currents from the rated maximum interrupting current down to the current that causes melting of the fusible element in one hour." (25°C, ambient)

This definition usually covers those currents that are between 150 to 200% of the fuse's rating. Although this ability includes a wide range of currents, it does not include all the possible currents that the fuse may be

109

called upon to clear. An example of this would be a fuse placed in a high ambient condition. It would be quite possible that a fuse could be required to interrupt a current much lower than the value stated by the one hour criteria. Some examples of general purpose fuses are:

RTE	-- ELS*
McGraw	-- Nx*
GE	-- Surge Guard GP
Westinghouse	-- CX

The backup fuse is capable of interrupting all currents in a particular range. It is defined by ANSI C37.40 - 1969 as follows:

"A fuse capable of interrupting all currents from the rated maximum interrupting current down to the rated minimum interrupting current (as specified by the manufacturer)."

Some examples of backup current limiting fuses are:

RTE	-- ELSP, ELO
McGraw	-- Companion (40k - only)
GE	-- OSP, ETP, EJO
S&C	-- Fault Fiter
Kearney	-- Type A
Westinghouse	-- Type CL
Chance	-- K Mate

The backup fuse is by design a fuse that clears only the high fault currents. The low current clearing is accomplished through the use of an expulsion fuse connected in series with the current limiting backup fuse. Its construction is therefore simpler. Although the element design is extremely important in controlling the let-through energy, no provisions are required for low-fault current clearing. When such clearing is required, the time-current characteristics are a composite of the two fuses as shown in Figure 4-8 below, and each fuse works in its optimum range.

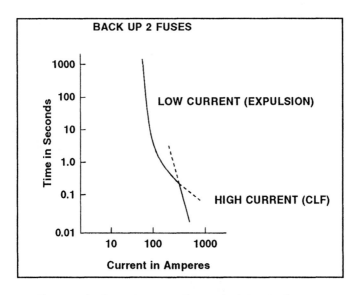

Figure 4-8. Time Current Characteristics for Backup Current Limiting Fuse

Coordination. The initial operating mechanism of a fuse is the melting of the element. This mechanism depends on the following three factors:

• Magnitude of current
• Duration of current
• Electrical properties of the element.

It is the time current curve (TCC) which defines the characteristic of a fuse. More specifically, the fuse characteristic is defined by 2 curves, the minimum melt (mm) and the total clearing (TC).

The minimum melt curve is developed by electrical test. The magnitude of the current and the time it took the fuse to melt are recorded and plotted. At this point, a curve is drawn through these points representing an "average" melt curve and from this curve, 10 percent is subtracted and the resulting curve is called the "minimum melt" curve.

The fuse, however, has an arcing time associated with it. The arcing time is the time it takes the fuse to interrupt the circuit after the fuse melts and it is also obtained by the test. The arc times, which are

111

recorded at different current magnitudes, are added to the "maximum melt" time (maximum melt time = 110 percent of average melt time). The resulting curve is called the "total clearing" curve. The two curves (see Figure 4-9 below), i.e., minimum melt and total clear, are the extremes of the fuse characteristics and are the curves published by each manufacturer.

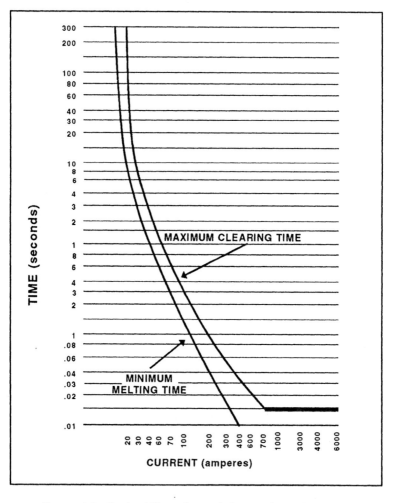

Figure 4-9. Typical Time Current Curves for a 10 K Line.

Relays

To some utilities, the distribution system starts at the primary of the distribution substation while other utilities classify their distribution system from the feeder breaker to the customer. In either case, a knowledge of overcurrent relaying is necessary.

Required Characteristics. The required characteristics necessary for protective equipment to perform its function properly are: sensitivity, selectivity and speed. This is especially true for relays.

Sensitivity: Sensitivity applies to the ability of the relay to operate reliably under the actual condition that produces the least operating tendency. For example, a time-overcurrent relay must operate under the minimum fault current condition expected. In the normal operation of a power system, generation is switched in and out to give the most economical power generation for different loads which can change at various times of the day and various seasons of the year. The relay on a distribution feeder must be sensitive enough to operate under the condition of minimum generation when a short circuit at a given point to be protected draws a minimum current through the relay. (NOTE: On many distribution systems, the fault-current magnitude does not differ very much for minimum and maximum generation conditions because most of the system impedance is in the transformer and lines rather than the generators themselves.)

Selectivity: Selectivity is the ability of the relay to differentiate between those conditions for which immediate action is required and those for which no action or a time-delayed operation is required. The relays must be able to recognize faults on their own protected equipment and ignore, in certain cases, all faults outside their protective area. It is the purpose of the relay to be selective in the sense that, for a given fault condition, the minimum number of devices operate to isolate the fault and interrupt service to the fewest customers possible. An example of an inherently selective scheme is differential relaying; other types, which operate with time delay for faults outside of the protected apparatus, are said to be relatively selective. If protective devices are of different operating characteristics, it is especially important that selectivity be established over the full range of short circuit current magnitudes.

113

Speed: Speed is the ability of the relay to operate in the required time period. Speed is important in clearing a fault since it has a direct bearing on the damage done by the short circuit current; thus, the ultimate goal of the protective equipment is to disconnect the faulty equipment as quickly as possible.

Characteristics of Overcurrent Relays. The overcurrent relay is the simplest type of protective relay (see Figure 4-10). As the name implies, the relay is designed to operate when more than a predetermined amount of current flows into a particular portion of the power system. There are two basic forms of overcurrent relays: the instantaneous type and the time-delay type.

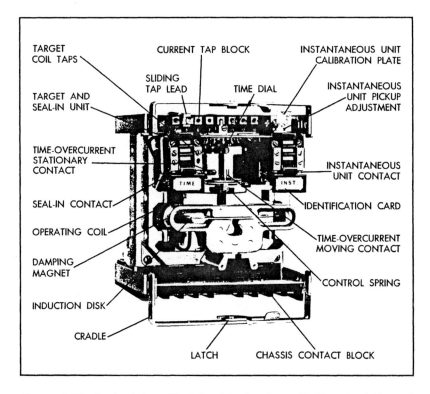

Figure 4-10. Typical Type 51 Relay Mechanism with Standard Hinged Armature Instantaneous Unit Withdrawn from Case

The instantaneous overcurrent relay is designed to operate with no intentional time delay when the current exceeds the relay setting. Nonetheless, the operating time of this type of relay can vary significantly. It may be as low as 0.016 seconds or as high as 0.4 seconds. The operating characteristic of this relay is illustrated by the instantaneous curve of Figure 4-11.

115

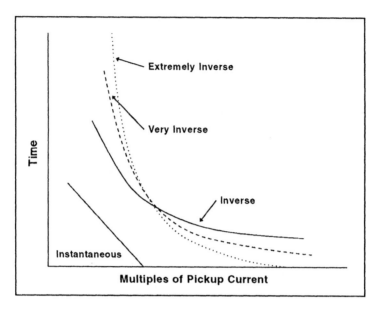

Figure 4-11. Time-Current Characteristics of Overcurrent Relays

The time-overcurrent relay (IAC, CO) has an operating characteristic such that its operating time varies inversely as the current flowing in the relay. This type of characteristic is also shown in Figure 4-11. The diagram shows the three most commonly used time-overcurrent characteristics: inverse, very inverse, and extremely inverse. These curves differ by the rate at which relay operating time decreases as the current increases.

Both types of overcurrent relays are inherently non-selective in that they can detect overcurrent conditions not only in their own protected equipment but also in adjoining equipment. However, in practice, selectivity between overcurrent relays protecting different system elements can be obtained on the basis of sensitivity (pickup) or operating time or a combination of both, depending on the relative time-current characteristics of the particular relays involved.

These methods of achieving selectivity will be illustrated later. Directional relays may also be used with overcurrent relays to achieve selectivity.

The application of overcurrent relays is generally more difficult and less permanent than that of any other type of relaying. This is because the operation of overcurrent relays is affected by variations in short-circuit-current magnitude caused by changes in system operation and configuration. Overcurrent relaying in one form or another has been used for relaying of all system components. It is now used primarily on distribution systems where low cost is an important factor.

Figure 4-12 shows a family of inverse time curves of the widely used IAC relay, which is an induction disc type. The time curves for the new design static overcurrent relays are similar.

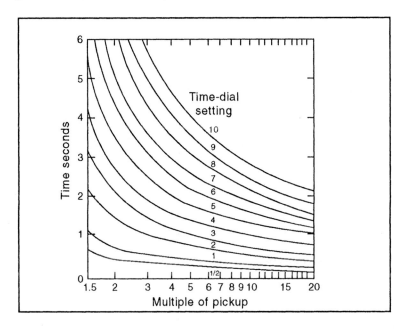

Figure 4-12. Inverse Time Curves

A curve is shown for each numerical setting of the time dial scale. Any intermediate curves can be obtained by interpolation since the adjustment is continuous.

It will be noted that the curves shown in Figure 4-13 are plotted in terms of multiples of pickup value, so that the same curves can be used for any value of pickup. This is possible with induction-type relays where the pickup adjustment is by coil taps, because the ampere-turns at

117

pickup are the same for each tap. Therefore at a given multiple of pickup, the coil ampere-turns, and hence the torque, are the same regardless of the tap used.

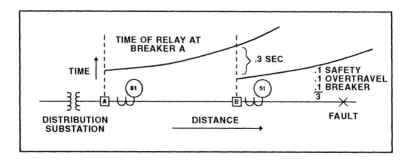

Figure 4-13. Operating Time of Overcurrent Relays With Inverse Time Characteristics

The time-current curves shown in Figure 4-12 can be used not only to determine how long it will take the relay to close its contacts at a given multiple of pickup and for any time adjustment, but also how far the relay disc will travel toward the contact-closed position within any time interval. For example, assume that the No. 5 time-dial adjustment is used and that the multiple of pickup is 3. It will take the relay 2.45 seconds to close its contacts. We see that in 1.45 seconds the relay would close its contacts if the No. 3 time-dial adjustment were used. In other words, in 1.45 seconds the disc travels a distance corresponding to 3.0 time-dial divisions, or three fifths of the total distance to close the contacts. For the most · effective use of an inverse-time relay characteristic, its pickup should be chosen so that the relay will be operating on the most inverse part of its time curve over the range of values of current for which the relay must operate. In other words, the minimum value of current for which the relay must operate should be at least 1.5 times pickup, but not very much more.

Figure 4-13 shows the application of time-overcurrent relays to a radial feeder and the total tripping time characteristics for faults at any location along a circuit. The figure shows the increase in the minimum tripping time as faults occur farther from the distribution substation - an increase inherent with overcurrent relaying. It also shows the effect of the inverse-time characteristic in reducing this increase. Obviously, the more line sections there are in series, the greater is the tripping time at the source end. It is not at all unusual for this time to be as high as 2 or 3 seconds. This is not a very long time according to some standards, but it would be intolerable if system stability or line burndown were an important consideration.

During light loads, some of the generators are usually shut down. At other times, the system may be split into several parts. In either case, the short circuit current tends to vary with the amount of generation feeding it. This generally does not have much effect at the distribution level. It should be appreciated that a reduction in the magnitude of short circuit current raises all of the characteristic curves of Figure 4-13. F o r locations where inverse time-overcurrent relays must be mutually selective, it is generally a good policy to use relays whose time-current curves have the same degree of inverseness. Otherwise, the problem of obtaining selectivity over wide ranges of short-circuit current may be difficult.

Instantaneous or undelayed overcurrent relaying is used only for primary relaying to supplement inverse-time relaying and is presently being used by most utilities. It can be used only when the current during short circuit is substantially greater than that under any other possible condition - for example, the momentary current that accompanies the energization of certain system components. The zone of protection of undelayed overcurrent relaying is established entirely by adjustment of sensitivity and is terminated short of the far end of the line. For instance, the instantaneous-overcurrent relay is usually set so that its pickup is 25 percent higher than the maximum current the relay will see for a three-phase fault at the end of the line. With this setting, the instantaneous relay will provide fault protection for about 80 percent of the line section.

Undelayed ("instantaneous") trips can frequently be added to inverse-time relaying and effect a considerable reduction in tripping time. This is shown in Figure 4-14 where the two sets of characteristics are superimposed. The time saved through the use of the instantaneous relays is shown by the shaded area. A reduction in the magnitude of short circuit current shortens the distance over which the instantaneous unit operates and may even reduce this distance to zero. However, this fact is usually of no great importance since faster tripping under the maximum short-circuit conditions is the primary objective.

119

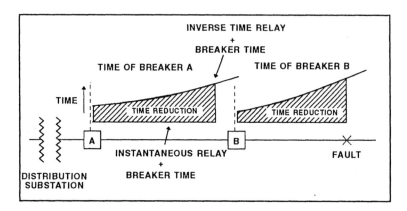

Figure 4-14. Reduction in Tripping Time Using Instantaneous Relaying

Instantaneous tripping is feasible only if there is a substantial increase in the magnitude of the short circuit current as the short circuit is moved from the far end of a line toward the relay location. This increase should be at least two or three times. For this reason, it often happens that instantaneous relaying can be used only on certain lines and not on others.

On systems where the magnitude of short circuit current flowing through any given relay is dependent mainly upon the location of the fault to the relay, and only slightly or not at all upon the generation in service, faster clearing can usually be obtained with very-inverse-time-overcurrent relays. Where the short circuit current magnitude is dependent largely upon system-generating capacity at the time of the fault, better results will be obtained with relays having inverse-time operating characteristics.

However, towards the ends of primary distribution circuits, fuses are sometimes used instead of relays and breakers. In the region where the transition occurs, it is frequently necessary to use overcurrent relays having extremely inverse characteristics to coordinate with the fuse characteristics.

The extremely inverse relay characteristic has also been found helpful, under certain conditions, in permitting a feeder to be returned to service after a prolonged outage. After such a feeder has been out of service for so long a period that the normal "off" period of all intermittent loads (such as furnaces, refrigerators, pumps, water heaters, etc.) has been exceeded, reclosing the feeder throws all of these loads on at once without the usual diversity. The total inrush current, also referred to as cold-load pickup, may be approximately four times the normal peak-load current. This current decays very slowly and can be as high as 1.5 times normal peak current after as much as three or four seconds. Only an extremely inverse characteristic relay provides selectivity between this inrush and short circuit current.

Whenever service has been interrupted to a distribution feeder for 20 minutes or more, it may be extremely difficult to re-energize the load without causing protective relays to operate. The reason for this is the flow of abnormally high inrush current resulting from the loss of load diversity. High inrush currents are caused by:

a. Magnetizing inrush currents to transformers and motors,
b. Current to raise the temperatures of lamp filaments and heater elements, and
c. Motor-starting current.

Figure 4-15 shows the inrush current for the first five seconds to a feeder which has been de-energized for 15 minutes. The inrush current, due to magnetizing iron and raising filament and heater elements temperatures, is very high but of such a short duration as to be no problem. However, motor-starting currents may cause the inrush current to remain sufficiently high to initiate operation of protective relays. The inrush current in Figure 4-15 is above 200 percent for almost two seconds.

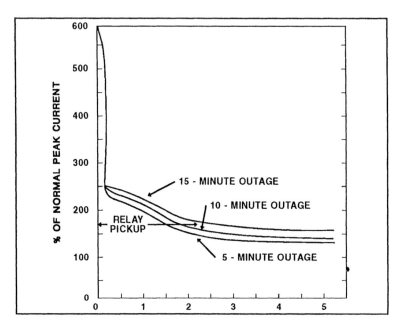

Figure 4-15. Five-, Ten-, and Fifteen-Minute Outage Pickup Curves for First Five Seconds after Restoral

The magnitude of cold load current is closely related to load diversity, but quite difficult to determine accurately because of the variation of load between feeders. If refrigerators and deep freeze units run five minutes out of every 20, then all diversity would be lost on outages exceeding 20 minutes.

A feeder relay setting of 200 to 400 percent of full load is considered reasonable. However, unless precautions are taken, this setting may be too low to prevent relay misoperation on inrush or cold load following an outage. Increasing this setting may restrict feeder coverage or prevent a reasonable setting of fuses and relays on the source side of this relay.

A satisfactory solution to this problem is the use of the extremely inverse relay. Figure 4-16 shows three overcurrent relays which will ride over cold-load inrush. However, the extremely inverse curve is superior in that substantially faster fault-clearing time is achieved at the high-current levels.

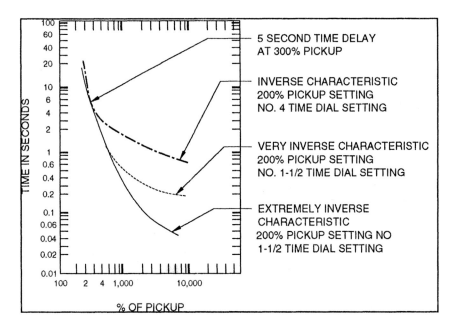

Figure 4-16. Comparison of Overcurrent Relay Characteristics

This figure, for the purpose of comparison, shows each characteristic with a pickup setting of 200 percent peak load and a five second time delay at 300 percent peak load to comply with the requirements for re-energizing feeders.

It is evident that the more inverse the characteristic, the more suitable the relay is for feeder short circuit protection. The relay operating time, and hence, the duration of the fault can be appreciably decreased by using a more inverse relay. Comparing the inverse characteristic shows that the extremely inverse characteristic gives from 30 cycles faster operation at high currents to as much as 70 cycles faster at lower currents.

Unfortunately, the extremely inverse relay may not always take care of the problem. As the feeder load grows, the relay pickup must be increased and a point may be reached at which the relay cannot detect all faults. At this time, it may be necessary to either move the fuses or reclosers closer to the substation or use automatic sectionalizing.

123

Reclosers

An automatic circuit recloser is a self-contained device which can sense and interrupt fault currents as well as re-close automatically in an attempt to re-energize a line. The operation of a recloser is very similar to that of a feeder breaker with a reclosing relay. The main difference between the two devices is that the recloser has less interrupting capability and costs considerably less.

Recloser operation utilizes two inverse time curves. The first curve, referred to as the "instantaneous" or "A" curve (see Figure 4-17) is similar to an instantaneous relay and is used primarily to save lateral fuses under temporary fault conditions. The second curve referred to as the "time delay" or "B" (also C, D, or E) is used to delay recloser tripping and allow the fuse to blow under permanent fault conditions. As can be seen in Figure 4-17, a 50 ampere recloser (hydraulic) has a minimum trip for both the instantaneous and time delay currents of 100 amperes or twice its rating.

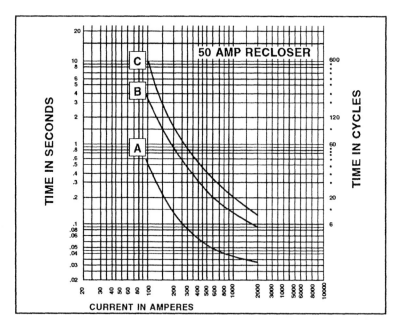

Figure 4-17

The reclosers "time-delay" curves (B, C, D, E) are fixed. This is in contrast to an inverse-relay curve which has an infinitely adjustable time dial. On the other hand, a recloser may have more than one instantaneous operation in an attempt to dissipate a temporary fault condition whereas a feeder relay can normally have only one. The most typical reclosing sequence for a recloser is two fast operations followed by two slow (2A, 2B). This is illustrated below in Figure 4-18.

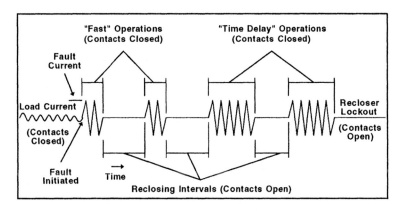

Figure 4-18. Typical Reclosing Sequence

Major classifying features in automatic circuit reclosers are:

- Single- or three-phase
- Control: hydraulic or electronic
- Interrupters: oil or vacuum.

Single- or Three-Phase. Single-phase reclosers are used to protect single-phase lines such as branches or taps of a three-phase. Also can be used to isolate single-phase loads.

Three-phase reclosers are used where lockout of all three phases is required for any permanent fault. They are also used to prevent single-phasing of three-phase loads such as large three-phase motors.

Controls: Hydraulic or Electronic. The intelligence that enables a recloser to sense overcurrents, select timing operation, time the tripping and reclosing functions, and finally lockout, is provided by its control.

There are two basic types of control schemes used: an integrated hydraulic control or an electronic control located in a separate cabinet.

Hydraulic recloser control is used on all single-phase reclosers and in smaller ratings of three-phase reclosers made by McGraw (GE did not use hydraulic controls). It is built as an integral part of the recloser. With this type of control, an overcurrent is sensed by a trip coil that is connected in series with the line. When the overcurrent flows through the coil, a plunger is drawn into the coil to trip open the recloser contacts. Minimum trip current for this type of recloser is two times the rating of the recloser.

The electronic control is more flexible, more easily adjusted and more accurate. The electronic control conveniently permits changing timing, trip current levels, and sequences of recloser operations without deenergizing or untanking the recloser. Line current is sensed by special CT's in the recloser. Minimum trip level is independent of recloser rating.

Interrupters - Oil, Vacuum, and SF$_6$. Reclosers using oil (GE and McGraw) for current interruption use the same oil for basic insulation. Vacuum (also supplied by GE and McGraw) provides the advantage of lower maintenance frequency. Depending on type, a vacuum recloser may use either oil or air as the basic insulating medium. The newly introduced recloser by A.B. Chance uses SF$_6$.

To properly apply automatic circuit reclosers, five major factors must be considered:

1. System voltage
2. Maximum available short circuit level
3. Maximum load
4. Minimum fault current (in protected zone)
5. Coordination with other devices.

Ratings of reclosers are generally considerably less than the ratings of popular feeder breakers in both continuous current and short circuit interruption capability. Units made by GE (now Multiamp) some years ago were limited to 4000 amperes and less as shown in Table 4-1.

Table 4-1. GE Recloser			
Frame Size Amps	kV Rating	Phase	Short Circuit Interrupting
50-280	14.4	1 & 3	125-4000 A
100	24.9	1 & 3	200-2500 A

A summary of McGraw's (now Cooper Power) capability is as follows:

Table 4-2. McGraw Recloser			
Frame Size	kV Rating	No. of Phases	Interrupting Amps
50-560	2.4-14.4	1	125-10,000
100	24.9	1	300-8,000
100-560	2.4-14.4	3	200-20,000
560	24.9	3	3,000-12,000
560	34.5	3	16,000

Sectionalizers

A sectionalizer is a protective device, used in conjunction with a recloser (or breaker with reclosing relay) which isolates faulted sections of lines. The sectionalizer does not interrupt fault current. Instead, it counts the number of operations of the reclosing device and opens while this backup device is open. After the sectionalizer opens, clearing the faulted section, the backup devices recloses to return power to the unfaulted sections of the line. If the fault is temporary, the sectionalizer will reset itself after a prescribed period of time.

A sectionalizer provides several advantages over fuse cutouts. In addition to application flexibility, they offer safety and convenience. After a permanent fault, the fault-closing capability of the sectionalizer greatly simplifies the testing of the circuit, and if the fault is still present, interruption takes place safely at the backup recloser.

Replacement fuse links are not required, thus the line can be tested and service restored with far more speed and convenience. Also, the possibility of error in the selection of the right size and type of fuse link is eliminated.

Because it has no time-current characteristic, a sectionalizer has distinct application advantages.

1. It can be applied between two protective devices having operating curves which are close together. This is a vital feature in a location where an additional step in coordination is not practical or possible.
2. It can be used on close-in taps where high fault magnitude prevents coordination with fuses.

Some of the disadvantages associated with sectionalizers are as follows:

1. <u>Cost</u> - The major reason utilities don't use more sectionalizers is that the advantages of the sectionalizer over a fuse are not justified by the cost differential of over $500. Also, sectionalizers have technical problems of their own.
2. <u>Memory Time</u> - Hydraulically controlled and dry type sectionalizers have some problems with memory time. In a standard sectionalizer, the reset time after a transient fault depends on the number of counts and the memory time selected. It can range from 5 to 22 minutes. Corresponding reset times for reclosers are from 10 to 180 seconds. Hydraulically controlled and dry type sectionalizers do not provide a choice of memory times. Memory time is essentially a function of the viscosity of oil which in turn is dependent on temperature. Consequently, accuracy predicting memory times is impossible and a too long memory time <u>may result in miscoordination during temporary faults</u> (as well as to a lesser degree permanent faults). Areas with high isokeraunic levels are particularly susceptible. The sectionalizer memory time must be sufficiently long such that the sectionalizer will retain its counts throughout the entire tripping and reclosing sequence of the backup fault interrupter. The memory time of hydraulic and dry-type sectionalizers vary with temperature, and this variable should be included in the calculation process. The consideration is not included here since the process is dependent on the type and manufacturer of the individual sectionalizer.
3. <u>Inrush</u> - Inrush has been a very big problem for some sectionalizers. The problem has been that the sectionalizer is very fast and sees inrush currents as fault currents. As such, it may misoperate (discussed in the next section on coordination).

COORDINATION OF DEVICES

Fuse-to-Fuse Coordination (Expulsion)

Expulsion Fuse to Expulsion Fuse. Figure 4-19 illustrates the general principle of coordinating expulsion fuses in series. Fuse (1) is called the protected fuse, and fuse (2) is called the protecting fuse. For perfect coordination, fuse (2) must melt and clear the fault before fuse (1) is damaged. To ensure this, three things are done:

1. The maximum characteristic (the total clear curve) of the protecting fuse (2) is plotted to conservatively estimate the maximum duration of the fault current.
2. The minimum (published) characteristic of the protected fuse (1) is plotted.
3. Seventy-five percent of the MM curve of the protected fuse is plotted to make sure that the fuse is not damaged and to account for any degradation in the fuse characteristics. (This is sometimes called the "damage curve.")

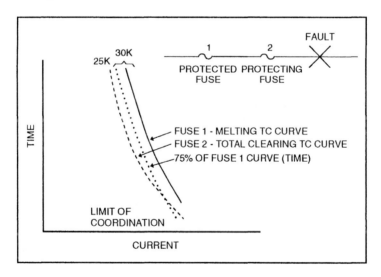

Figure 4-19. Coordination of "K" Links

If the damage curve of fuse (1) and the TC curve of fuse (2) never cross, there is said to be perfect coordination. If they cross, however, at some value of current, this is called the "limit of coordination." For example, suppose the curves cross at 2000 amperes. This means that coordination above 2000 amperes is unlikely. However, if the maximum available short circuit of the system at that location is only 1500 amperes the fuses would be considered to be fully coordinated.

Current Limiting Fuse to Expulsion Fuse. The basic problem in using a full range current limiting fuses is that their normal TC characteristics are different from most other devices making them difficult to coordinate.

A situation which commonly exists for a utility is the use of an expulsion fuse at a lateral tap and a CLF at a transformer as shown below.

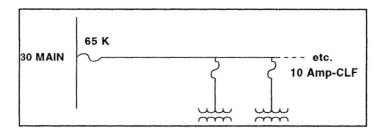

Figure 4-20

For a fault on the transformer, we want the CLF to clear the fault without damage to the 65 K fuse (protected). A plot of the fuse characteristics using the 75 percent rule (i.e., 4 melt on 3 clear) is shown in Figure 4-21.

As can be seen, these two fuses coordinate quite well at least down to .01 seconds. As we know, however, the CLF is a current limiting device and can in effect force a current zero which in effect means that the CLF can melt in less than .01 seconds. This energy limiting characteristic actually protects the K link and full coordination down to 350 amperes is assured.

Another check on this ability to coordinate in less than .01 seconds can be illustrated by comparing the minimum melt (MM) I^2t of the 65 K link to the total clearing (TC) I^2t of the 10 ampere CLF. For example, the total maximum I^2t of a 10 amp CLF is less than 4400 amp^2 sec at any

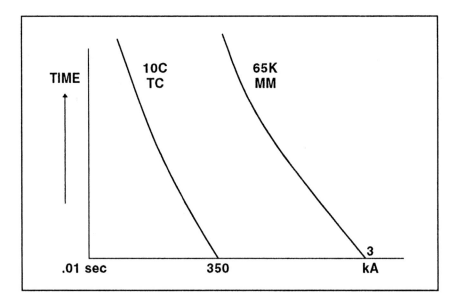

Figure 4-21

voltage rating (per the manufacturer) whereas the minimum melt I^2t of the expulsion is a calculated value at the .01 sec point, i.e., $(3000)^2(.01)$ = 90,000 amp^2 sec. It is evident, therefore, that the CLF will always melt well before the K link and coordination up to 50,000 amperes is assured.

On the other hand, when the CLF is the protected fuse and the expulsion is the protecting fuse, coordination is limited. A lateral protected by a CLF is shown below in Figure 4-22.

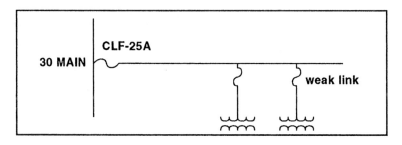

Figure 4-22

131

The reason for the limited coordination is that the expulsion fuse must wait for a current zero to interrupt. As a result, an asymmetric current may flow in the device for up to .013 seconds. If we plot these fuses (see Figure 4-23) we can see that coordination exists up to only 500 amperes. Above this fairly low level, it is probable both fuses would operate.

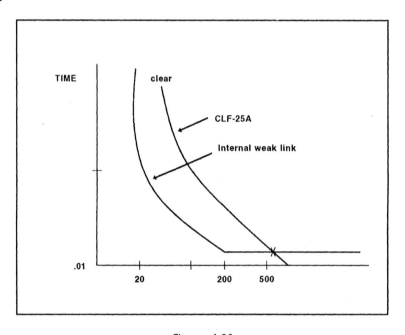

Figure 4-23

Current Limiting Fuse to Current Limiting Fuse. When a current limiting fuse is used both as the "protecting" and "protected" fuse, we expect to see a coordination plot, now with fuses of similar time-current characteristics.

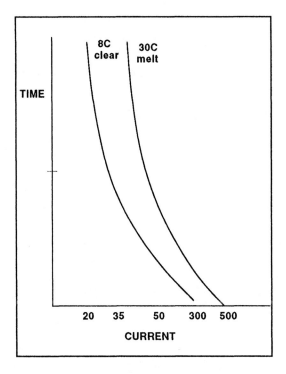

Figure 4-24

Since current limiting fuses are current limiting devices and can operate in less than .01 seconds, additional check is needed to make sure coordination below .01 seconds takes place. All manufacturers of current limiting fuses <u>determine from tests</u> and publish minimum melt I^2t and total maximum I^2t values for the purpose of coordination.

Listed below are the minimum melt as I^2t and total let through I^2t values for general purpose current limiting fuses rated 15.5 kV.

133

Table 4-3. CLF Characteristics		
Fuse Rating Amps	**I^2t Min. Melt**	**I^2t Max. Total**
6C	150	1280
8C	230	2500
10C	520	3200
12C	1160	9800
15C	1540	12000
20C	2690	16500
25C	4560	25000
30C	4560	16000
40C	10700	40000

For the 8C and 30C fuses in the coordination plot, we see that the MM I^2t of the 30C fuse is almost twice the max total I^2t of the 8C protecting fuse. Consequently, coordination exists. Table 4-3 illustrates how a large fuse may not coordinate with a small rated fuse even though they have similar characteristics. For example, any "protecting fuse" above 10 amperes will not coordinate with the 30C fuse used in this example.

Backup Current Limiting Fuse Coordination - "Two Fuse System". A backup current limiting fuse is always applied with a series-connected expulsion fuse. The procedure mandates coordinating the time current characteristics so the expulsion fuse interrupts all currents below the rated minimum interrupting current of the backup fuse. Two coordination choices are possible depending upon the characteristics of the backup fuse:

- Crossover coordination
- Matched melt coordination

The expulsion fuse is chosen in the conventional manner based on inrush considerations and overload protection practice, if there are no supplementary secondary protective devices. With primary fusing only, protection against overloading or low fault currents in the region of 2 to 4 times full nameplate load current is most commonly used. When secondary breakers of fuses are involved, the primary expulsion fuse must be coordinated with these devices.

Crossover Coordination. Figure 4-25 illustrates both the crossover and minimum melt I^2t approaches.

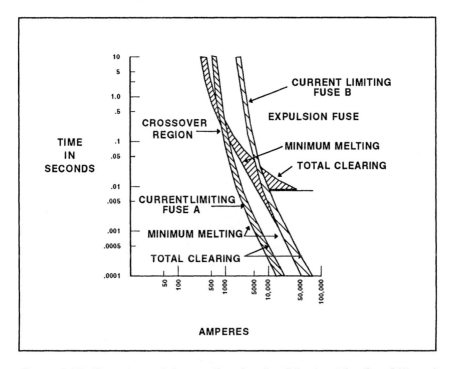

Figure 4-25. Time-Current Curves Showing the Effects of the Two Different Crossover Regions Currently in Use

Matching the expulsion fuse with CLF "A" involves a crossover of characteristics in the time region between .01 and 100 seconds. This is an approach sometimes preferred as an oil submersible protection (OSP) where the CLF is under oil in the transformer and the fuse may be of a bayonet type. In this case it may be desirable to have the CLF operate only for very high fault currents that might be seen in a transformer failure where the unit must be removed anyway. Secondary faults of lower magnitude would be interrupted by the easily replaceable expulsion fuse. For successful coordination, the backup CLF must have a minimum interrupting capability which includes the diamond shaped crossover region (plus some safety factor). In the case of fuse A in the illustration,

135

this would require a minimum current interrupting capability of approximately 700 amperes.

Matched Melt Coordination. This coordination method requires that the minimum melt I^2t of the expulsion fuse be equal to or less than the backup current limiting fuses at .01 seconds. This, in fact, means that the expulsion fuse will melt and arc for all levels of fault current and will share the interrupting duty. Also, the interruption of the expulsion fuse will take voltage from the CLF, allowing simpler design of the backup CLF. The relative amount of interrupting duty that will be imposed upon the expulsion fuse will depend on how early it melts on the current wave. A comparison of the melting I^2t for the expulsion fuse versus the backup current limiting fuse will indicate relative melting times. Fuses must be matched carefully since too large an expulsion fuse will not properly protect the current limiting fuse from fault currents below its interrupting capability, and too small an expulsion fuse may impose too much of the interrupting duty on the expulsion fuse. This latter factor is of less importance for expulsion fuses mounted in cutouts, but certain types of internal transformer expulsion fuses have a limited energy withstand capability. The combination of the expulsion fuse and the series connected backup current limiting fuse should be tested to assure proper interrupting performance.

The total clearing I^2t for the backup current limiting fuse used in this approach (fuse B in the illustration above) will be larger than that of fuse A, even though it matches the same expulsion fuse. For many transformer voltage and kVA ratings, this higher let-through current should not be a problem but testing will be required to verify this.

If the expulsion fuse of approach 2 is an internal transformer fuse, its arcing will add some energy to that generated by the transformer internal arcing fault. Due to the presence of the current limiting fuse, this energy increase will be significantly less than it would otherwise be. The resultant duty on the transformer is therefore increased somewhat as compared with approach 1 (fuse A).

In order to assure proper coordination between fuses, manufacturers prepare tables illustrating limits for coordination between expulsion and backup CLFs. Shown in Table 4-4 is a typical table put together for use with externally applied backup CLFs.

Likewise, for an oil submersible (OS) fuse a table similar to the one shown in Table 4-5 is prepared by each manufacturer.

Table 4-4. Fuse Selection Data Summary

Fuse Rating			Application Information					Recommended Maximum Size of Frequently Used Series-Connected Expulsion Fuses		
Max Design Volts (kV)	K-Link Coordination	System Voltage Class (kV)	Peak Arc Volts (kV)	Maximum Continuous Current (Amps)	Minimum Interrupting Current (Amps)	Minimum Melt i^2t (A²SEC)	Maximum Total i^2t (A²SEC)	K	T	QA
8.3	12	15	26	20	415	3,200	10,000	12	8	15
8.3	25	15	26	40	500	11,000	33,000	25	15	30
8.3	40	15	26	70	750	28,000	80,000	40	20	50
8.3	50	15	26	80	850	39,000	120,000	50	25	60
15.5	12	27	49	20	440	3,200	10,000	12	8	15
15.5	25	27	49	40	580	11,000	33,000	25	15	30
15.5	40	27	49	65	850	28,000	80,000	40	20	50
15.5	50	27	49	75	1000	39,000	120,000	50	25	60
23.0	12	35	64	20	280	3,200	10,500	12	8	15
23.0	25	35	66	40	465	11,000	38,000	25	15	30

NOTE: All designs have a 50,000 amps rms symmetrical rating

Table 4-5. Coordination Tables for Expulsion Fuse/OS Fuse

Single Phase - Bay-O-Net Dual (Load) Sensing RTE Line
358C, GE Link 9F59LFC/Minimum Acceptable OS Fuse

	OS Fuse Voltage Rating (kV)											
	2.4		8.3				15.5				23.0	
	Transformer Voltage (kV)											
	2.4		4.16-4.8		7.2-7.96		12-12.47		13.2-14.4		19.9	
KVA	Link	OS	Link	OS	Link	OS	Link	OS	Link	OS	Link	OS
5	C03	40	C03	40	C03	40	C03	40	C03	40	C03	40
10	C05	40	C05	40	C03	40	C03	40	C03	40	C03	40
15	C08	50	C05	40	C03	40	C03	40	C03	40	C03	40
25	C10	80	C08	50	C08	40	C05	40	C03	40	C03	40
37.5	C12	125	C10	80	C08	40	C05	40	C05	40	C03	40
50	C12	150	C10	80	C08	50	C05	40	C05	40	C05	40
75	-	-	C12	150	C10	80	C08	50	C08	50	C05	40
100	-	-	C12	200	C10	100	C08	50	C08	50	C05	40
167	-	-	-	-	C12	200	C10	100	C10	100	C08	50
250	-	-	-	-	-	-	C12	250	C12	150	-	-
333	-	-	-	-	-	-	C12	250	C12	250	-	-

Distribution Transformer Fusing

Purpose of the Fuse. With reference to Figure 4-26, which shows a transformer connected phase-to-neutral in a single-phase circuit of a multigrounded system, the basic functions of a fuse are:

a. Isolate a transformer with an internal fault from the primary circuit so that only those customers served from the faulted transformer experience a service interruption. Ideally this isolation process will be accomplished such that the transformer does not fail in a disruptive fashion.

b. Protect the transformer against the effects of through fault currents from bolted faults at and downstream from the secondary (low-voltage) terminals of the transformer. The fuse should operate before the transformer is damaged thermally or mechanically.

c. Protect, to whatever extent possible, the transformer for high impedance faults or arcing faults in the secondary circuits fed from the transformer.

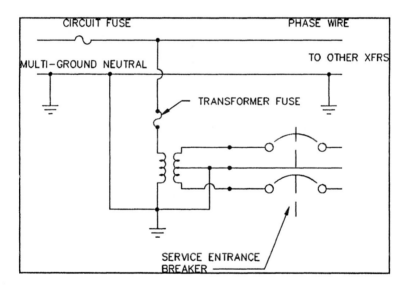

Figure 4-26. Single-Phase Circuit With Line-to-Neutral Connected Distribution Transformer

In addition to performing the functions given by items a, b, and c above, the fuse applied with the distribution transformer must satisfy overcurrent coordination criteria. The two principal coordination criteria which the transformer primary fuse should satisfy are:

a. Coordinate selectivity with the upstream line fuse, line recloser, or circuit breaker such that only the transformer fuse blows for a permanent fault downstream from the transformer fuse, thereby preventing blowing of the line fuse or operation of the recloser or breaker on their time delay curves.
b. Coordinate selectively with the first downstream overcurrent protective device in the secondary system, which frequently is a main breaker or fuse at the service entrance equipment or panel.

And finally, the fuse should not operate for any conditions, such as short time overloads or transient currents, which are not harmful to the transformer or connected secondary system. The main conditions which the fuse should withstand without operating (melting) are:

a. Transient surge currents through the fuse due to lightning, magnetizing inrush current due to core saturation, and cold load pickup currents following an extended outage.
b. Short time overload currents which are within the thermal capacity of the transformer.

Inrush/Cold Load. Minimum transformer fuse size is usually determined by the protection of the following three points:

a. One "cold load pickup" point (3 times full load for 10 seconds)
b. Two "inrush" points:
         (12 times full load for 0.1 second)
         (25 times full load for 0.01 seconds)

As an example you are asked to protect a transformer rated 500 kVA, 34,500/19,000 with "K" type fuses. Considering inrush and cold load pickup, what is the smallest fuse you can use (neglect damage curve of fuse)?

Solution:

a.   Full Load Current

$$= \frac{kVA}{\sqrt{3} \cdot E_{LL}}$$

4-1

$$= \frac{500}{\sqrt{3} \cdot 34.5 \ kV} = 8.4 \ \text{amps}$$

b.   Inrush Criteria

$$12 \times 8.4 = 100.8 \ \text{amps at} \ .1$$

4-2

$$25 \times 8.4 = 210 \ \text{amps at} \ .01$$

c.   Cold Load

$$3 \times 8.4 = 25.2 \ \text{amps at} \ 10$$

4-3

Plotting the points on a "K" link fuse curve would suggest a fuse size of at least 12K is required.

Twice the Load. Another way to select the minimum fuse size (used by the author) is to simply use a fuse size approximately 2 times the full load current of the transformer. For the previous example, we would select a 20K load based on the calculation ($2 \times 8.4 = 16.8$ amperes). The disadvantage to this higher rating is that some secondary protection is lost and there is no overload protection. The arguments in favor of this approach are:

a.   It greatly reduces the number of nuisance fuse blowings caused by multiple inrush during reclosing.
b.   It reduces nuisance fuse blowings due to saturation of the transformer caused by lightning.
c.   Inrush can be higher than previously thought.
d.   Overload protection is not the purpose of the fuse.
e.   Secondary protection is not the purpose of the fuse.
f.   It's easy!

Capacitor Fusing

Capacitors should be fused as close as possible to minimize the chance of case rupture but not to close so as to create a problem with nuisance fuse blowings. The standards recommend fusing to at least 135% of load (some manufacturers recommend 165%). It must be kept in mind that some fuses have continuous overload capability.

Problem. Choose an expulsion fuse for a capacitor bank rated 150 kVAR, 3-phase, on a multigrounded 13.2/7.6 kV system. Assume the fuse has a continuous overload capability of 150%.

Solution.

$$\text{kVAR per phase} = \frac{150}{3} = 50$$

$$I = \frac{\text{kVAR}}{\text{kV}} = \frac{50}{7.6} = 6.56 \text{ amps} \qquad 4\text{-}4$$

Fuse size $* 1.5 \geq 6.56 * 1.35 \Rightarrow$
Fuse must be ≥ 5.9 amps

(choose a 6 amp fuse)

Setting an Overcurrent Relay

Setting an overcurrent relay is not particularly difficult, although it can be confusing. The main two items to be concerned with are that the relay does not trip at too low a current so that false trips occur and that the relay coordinates with the downstream device. The following two examples demonstrate some of these considerations. The second example coordinates two relays in series on a distribution feeder. While this scenario is unusual, it is very similar to the case of having a recloser downstream and does demonstrate the same exact basic principle of setting the substation relay with enough safety margin to minimize false operation.

Example: Setting an IAC Relay. Assume that the IAC relay should trip on sustained current at 450 amperes minimum and 3750 amps in 1.9 seconds. Assume: CT is 60:1. In selecting an IAC relay, you must first select a tap and then select a time dial:

Step 1 - Current Tap

$$\text{Minimum Tap} = \frac{450}{60} = 7.5 \qquad\qquad 4\text{-}5$$

Since there is no 7.5 amp tap, use the 8 amp tap.

Step 2 - Time Dial. Need 3750 amps @ 1.9 seconds.

$$\frac{3750}{60} = 62.5 \text{ amps}$$

$$\therefore \frac{62.5}{8} = 7.8 \text{ multiples of pickup} \qquad\qquad 4\text{-}6$$

Example: Setting Two Devices (Breakers/Reclosers) in Series. The following is a simplified methodology to set devices in series:

1. Calculate short circuit levels

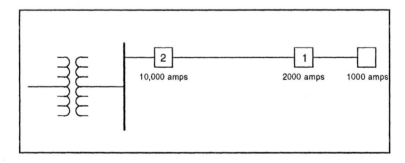

Figure 4-27. Distribution One-Line

143

2. Calculate relay minimum pickup:

$$\text{Relay } 1 = \frac{1000 \text{ amps}}{1.5} = 667 \text{ amps pickup}$$

4-7

$$\text{Relay } 2 = \frac{2000 \text{ amps}}{1.5} = 1330 \text{ amps pickup}$$

3. Assume margins of .3 sec. Time of relay 1, at a #1 T.D. (using Figure 4-28), for fault at breaker 1 line is found as follows:

$$\frac{2000 \text{ amps}}{667 \text{ amps}} = 3.0 \text{ times pickup}$$

4-8

$$\approx .2 \text{ seconds (on #1 T.D.)}$$

Relay 2 picks up at 1330 amps so the multiple of pickup for a fault at breaker location 1 is:

$$\frac{2000 \text{ amps}}{1330 \text{ amps}} = 1.5 \text{ multiples of pickup}$$

4-9

∴ The time is .5 (.2 + .3) seconds (see Figure 4-29) and the multiple of pickup is 1.5. Pick a #1 time dial (or greater).

Relay Curves

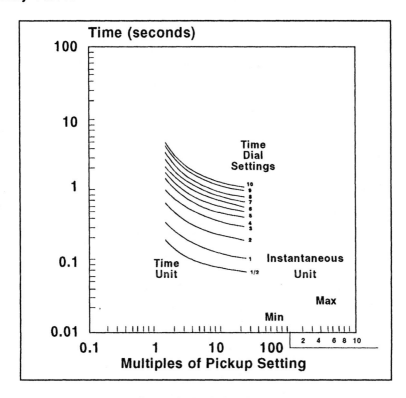

Figure 4-28. Relay Curve

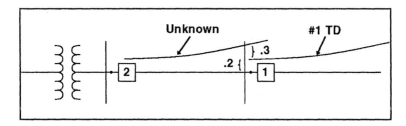

Figure 4-29. Safety Factor

Relay-to-Fuse (Feeder Selective Relaying)

The most common philosophy of feeder protection is to use "feeder selective relaying" which means that the feeder breaker and lateral fuse are coordinated in such a way that the lateral fuse only operates for permanent faults on the lateral. To accomplish this, the feeder breaker must operate before the fuse is damaged, as shown in Figure 4-30.

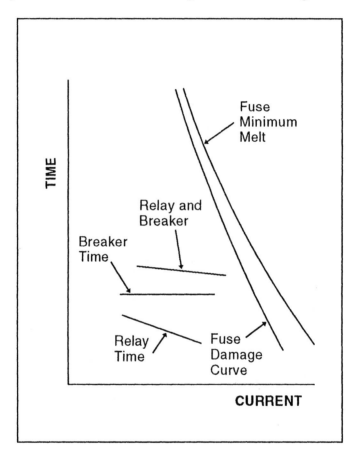

Figure 4-30. Coordination During Temporary Faults

Because the fuse is very fast at relatively high levels of short circuit current, it is sometimes impossible for the breaker to beat the fuse and consequently both devices will operate. For example the limits of coordination for various types of fuses (i.e., the highest current at which coordination can be expected), assuming a 6 cycle response of the relay and breaker, is as follows:

Table 4-6. Max Current at Which Coordination Is Possible	
Fuse Size	**Coordination Amperes**
100K	1200
100T	2000
200K	3500
200T	5800

For permanent faults, the fuse is expected to operate before the relay disc completes its travel as shown in Figure 4-31. The most common error when providing this type of coordination is to forget to consider the overtravel of the relay disc (assuming that an electromechanical relay is being used).

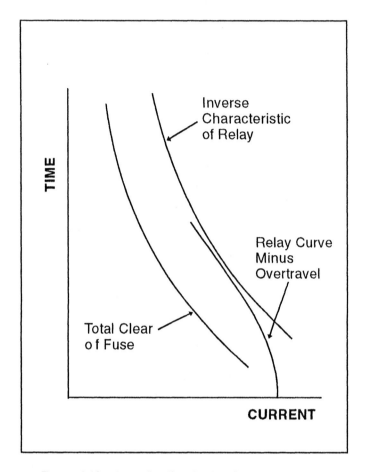

Figure 4-31. Coordination During Permanent Faults

While the theory of "feeder selective relaying" sounds good, it is difficult to implement because true coordination is limited to a very narrow range of fault currents. Figure 4-32 shows a very simplified illustration of why this is. As can be seen, for very low levels of currents the fuse may not operate as it is supposed to do for permanent faults. On the other hand, the fuse is too fast for high currents and will always operate. This would be a problem for temporary fault conditions.

On a distribution feeder, it is conceivable that all three conditions could exist, i.e., there are areas where the fuse always operates, never operates, and operates properly. This situation is shown in Figure 4-33. More realistically, the condition where the breaker or the recloser is always too fast for the fuse rarely occurs.

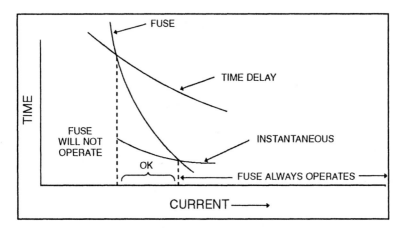

Figure 4-32.

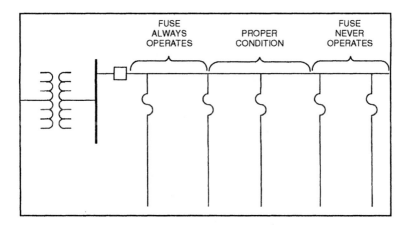

Figure 4-33.

Self-Extinguishing Lightning. The reclosing sequence of a feeder breaker is usually one fast trip followed by several time delayed trips. The time between the reclosures, i.e., the breaker is open, is called the "dead time". A typical sequence of dead times is 0,15,30 seconds. This is

149

shown in Figure 4-34. The instantaneous tripping takes about 6 cycles, which includes 1 cycle for the instantaneous relay and 5 cycles for the breaker. Some utilities in areas of high lightning activity have found that some lightning hits to the line are self-extinguishing. What this means is that after the lightning hits the line and initiates fault current (flashover) the fault may sometimes go out by itself. In most instances where this is successful, a utility has wooden crossarms. Wood has been found to possess good arc quenching capabilities. This being the case, some of these utilities will delay the instantaneous relay, which is about a 1 cycle device, by a few cycles thereby eliminating a needless trip of the feeder.

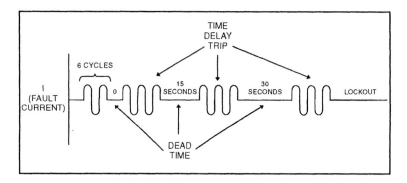

Figure 4-34. Feeder Breaker Reclosing

Elimination of Instantaneous Reclose. An instantaneous reclose means that upon its first trip, the breaker is immediately closed back into the circuit, i.e., no intentional time delay is introduced to the operation. The 0 second time shown in Figure 4-34 denotes such an instantaneous reclose. In reality, the duration of the instantaneous reclose is about 20 to 30 cycles due to the inertia of the breaker contact heads. Some utilities have found that the instantaneous reclosure is usually unsuccessful and the temporary fault reinitiates upon reclosure. This fault is successfully cleared after the first time delay trip where dead time is usually 5 seconds or more. The explanation given for the lack of success with instantaneous reclosing is that the ionized gases formed during the fault do not get a chance to dissipate if reclosure occurs too quickly. This is especially a concern for higher voltage systems using compact designs. Some of these utilities have found that by introducing a time delay of 2 or 3 seconds (up to 15 seconds) into the first dead time, the reinitiation can be prevented.

Power Quality. For a temporary fault on the lateral, a utility using "feeder selective relaying" will expect the feeder breaker to open, clearing the fault. The lateral fuse in this scenario will not be affected. The problem with this technique is that the entire feeder sees a momentary interruption and the "blinking clock syndrome" is created. In an effort to reduce the number of momentaries a customer sees and increase his so-called "power quality", many utilities are eliminating the instantaneous trip from feeder breaker. This, of course, means that temporary faults on the lateral now become permanent outages and as such are negatively reflected in reliability indices, i.e., average customer minutes out per year will increase.

High/Low Scheme. One philosophy the author has never seen used but suggests should be considered is having two philosophies of overcurrent protection on the same feeder, which we'll call the high/low scheme. Most feeders like the one shown below have areas of high fault current and areas of low fault current. It is suggested that the feeder breaker just protect the area of high fault currents. Because the fuse is faster than the breaker in this area, no instantaneous trip should be used since the fuse will operate anyway and the feeder trip will only cause more blinking clocks.

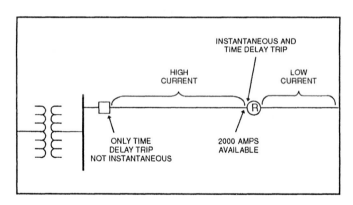

Figure 4-35. High/Low Scheme

151

It is suggested that for areas of the system where the fault currents are approximately 2000 amperes or less and coordination is possible with 100 amperes* or smaller fuses, reclosers be utilized and selective coordination (fuse only operates for permanent faults) be restored. This scheme, although requiring the addition of a recloser, reduces the number of momentaries as well as average customer minutes outages.

Recloser-to-Fuse (Lateral)

There is no real philosophy when it comes to lateral fusing but a few trends and observations can be mentioned:

Standardized Fuse Size. Most utilities pick a fuse size, like 65K, and use it for virtually all their lateral taps. The reason given is that it is easier for the crews to deal with one size. What it also says is the lateral loads are not really important and coordination is spotty. For example, Figure 4-36 shows a recloser/fuse coordination example. The coordination plot shows that total coordination exists only for points between "a" and "b", which are fault current levels. If we consider the one-line diagram and assume that fault levels "a" and "b" occur on the middle lateral, then we can conclude that laterals closer to and farther from this point will not totally coordinate. That is, since the fuse is the same size, the coordination plot and hence the limits stay the same. This technique, however, is probably as valid as any.

---

*Use of a 200 ampere fuse would shift the recloser location to a point on the system with a short current available of about 4000 amps.

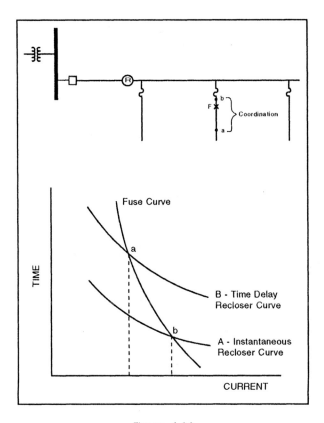

Figure 4-36.

Load Level. Some utilities fuse laterals depending on load. You can sometimes tell this because lateral fuses vary considerably and size is not a function of short circuit level (higher fuse sizes near the substation might indicate an attempt to coordinate). Fuses used for laterals should be rated at least twice the lateral load to allow for cold load pickup, inrush, and emergency backfeed. It should be noted that fusing of laterals does little if anything to prevent overload. The fusing philosophy, here, is to take out the fault, not protect for overload. Where lateral fuse sizes are still less than 25K or 15T, there is often a problem with lightning current blowing the fuse. Most lateral fuse operations during lightning storms, however, are caused by line flashover (fault current) which will operate any fuse size.

153

Fusing a lateral on the basis of load makes little sense, in the author's opinion, unless there is a conductor burndown problem. This should be a rare consideration on the newer designs. On some of the older construction, using small conductors, there is little choice but to fuse as tightly as possible and this philosophy is probably the best.

Feeder Selective Relaying (FSR)/Coordination. Some utilities pick a fuse size to allow maximum coordination with the breaker or recloser. Utilities using 100 or 200 ampere lateral fuses might possibly not be doing it because of lateral loading but rather because short circuit levels are relatively high and that's the only way to slow the fuse down enough on temporary faults to allow the breaker to operate.

Reclosers and Fuses. Figure 4-37 shows the time-current characteristic curves of the automatic circuit recloser. To these curves, the time-current characteristics of a fuse C is superimposed. It will be noted that fuse curve C is made up of two parts, i.e., the upper portion of the curve (low current range) represents the total clearing time T.C. curve, and the lower portion (high current range) represents the melting T.C. curve for the fuse. The intersection points of the fuse curves with the recloser curves "A" and "B" illustrate the limits between which coordination will be expected. Basically, this is correct within the interest of simplicity. However, to accurately establish intersection points "a" and "b", it is necessary that the characteristic curves of both recloser and fuse be shifted, or modified, to take into account alternate heating and cooling of the fusible element as the recloser goes through its sequence of operations. For example, since we want to protect the fuse for two instantaneous openings, as shown in Figure 4-38, it is necessary to compare the heat input to the fuse during these two instantaneous recloser openings.

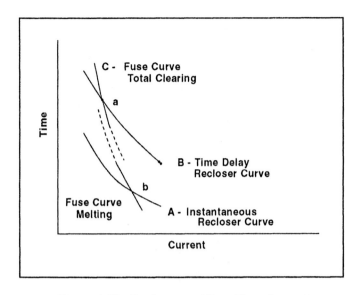

Figure 4-37. Recloser and Fuse Time-Current Characteristics

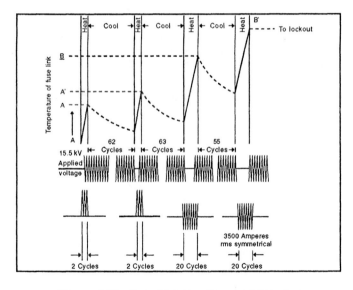

Figure 4-38. Fuse Link Heating and Cooling

One method used to account for the heating of the fuse is to shift the instantaneous curve to the right. For example, if the setting of the recloser was two instantaneous shots followed by two time delays and the "dead time" between reclosings was instantaneous, then the "A" curve would simply be doubled. Since there are usually about 2 to 5 seconds before reclosure, the fuse gets some chance to cool. To account for this, a factor less than two is used.

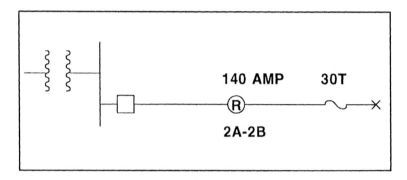

Figure 4-39. Recloser/Fuse Coordination

If we assume the system shown above and give the recloser a 2 second "dead time", we would shift the instantaneous curve by a factor of 1.35 and have the coordination plot shown in Figure 4-40. The limit of coordination would now be about 1000 amperes for faraway faults and over 5000 amperes for closed-in faults. If the system beyond the 30T fuse is within these limits, there is total coordination.

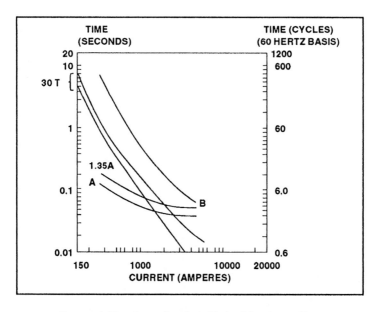

Figure 4-40. Coordination Plot of Recloser/Fuse

Relay-to-Recloser

If a permanent fault occurs anywhere on the system beyond a feeder, the recloser device will operate once, twice, or three times instantaneously (depending upon adjustment) in an attempt to clear the fault. However, since a permanent fault will still be on the line at the end of these instantaneous operations, it must be cleared away by some other means. For this reason, the recloser is provided with one-, two-, or three-time delay operations (depending upon adjustment). These additional operations are purposely slower to provide coordination with fuses or allow the fault to "self-clear". After the fourth opening, if the fault is still on the line, the recloser will lock open.

Figure 4-41 represents the instantaneous time delay characteristics of a conventional automatic circuit recloser.

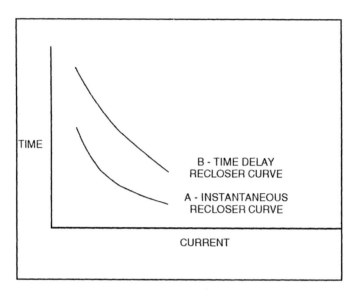

*Figure 4-41. Tripping Characteristic for Conventional
Automatic Circuit Recloser*

At substations where the available short circuit current at the distribution feeder bus is 250 MVA or more, the feeder circuits are usually provided with circuit breakers and extremely inverse-time overcurrent relays. The relays of each feeder should be adjusted so that they can protect the circuit to a point beyond the first recloser in the main feeder but with enough time delay to be selective with the recloser during any or all of the operations within the complete recloser cycle.

An important factor in obtaining this selectivity is the reset time of the overcurrent relays. If, having started to operate when a fault occurs beyond the recloser, an overcurrent relay does not have time to completely reset after the recloser trips and before it recloses (an interval of approximately one second), the relay may "inch" its way toward tripping during successive recloser operations. Thus it can be seen that it is not sufficient merely to make the relay time only slightly longer than the recloser time.

It is a good "rule of thumb" that there will be a possible lack of selectivity if the operating time of the relay at any current is less than twice the time delay characteristic of the recloser. The basis of this rule, and the method of calculating the selectivity, will become evident by considering an example.

First, it should be known how to use available data for calculating the relay response under conditions of possibly incomplete resetting. The angular velocity of the rotor of an inverse-time relay for a given multiple of pickup current is substantially constant throughout the travel from the reset (i.e., completely open) position to the closed position where the contacts close. Therefore, if it is known (from the time-current curves) how long it takes a relay to close its contacts at a given multiple of pickup and with a given time-dial adjustment, it can be estimated what portion of the total travel toward the contact-closed position the rotor will move in any given time. Similarly, the resetting velocity of the relay rotor is substantially constant throughout its travel. If the reset time from the contact-closed position is known for any given time delay adjustment, the reset time for any portion of the total travel (when the longest time delay adjustment is used) is generally given for each type of relay. The reset time for the number 10 time-dial setting is approximately six seconds for an inverse type overcurrent relay, and approximately 60 seconds for either a very inverse or any extremely inverse type overcurrent relay.

The foregoing information may be applied to an example by referring to Figure 4-42. Curves A and B are the upper curves of the band of variation for the instantaneous and time-delay characteristics of a 35 ampere recloser. Curve C is the time-current curve of the very inverse Type IAC relay set on the number 1.0 time-dial adjustment and 4 ampere tap (160 ampere primary with 200/5 current transformers). Assume that it is desired to check the selectivity for a fault current of 500 amperes. It is assumed that the fault will persist through all of the reclosures. To be selective, the IAC relay must not trip its breaker for a fault beyond the recloser.

The operating times of the relay and recloser for this example are:

Recloser:
 Instantaneous - 0.036 sec.
 Time delay - 0.25 sec.
Relay:
 Pickup - 0.65 sec.
 Reset - $(1.0/10)$ $(60) = 6.0$ sec.

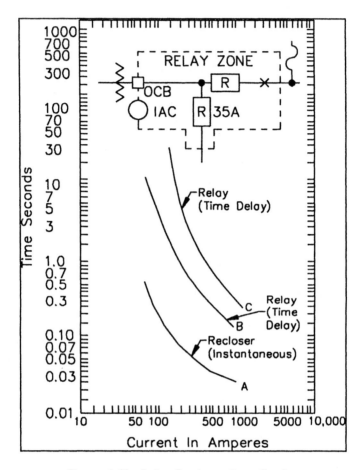

Figure 4-42. Relay-Recloser Coordination

The percent of total travel of the IAC relay during the various recloser operations is as follows, where plus means travel in the contact-closing direction and minus means travel in the reset direction:

Recloser Operation	Percent of Travel Relay Travel
First Instantaneous Trip	(0.036/0.65) x (100) = +5.5%
Open for One Second	(1/6) x (100) = -16.7%

It is apparent from this that the IAC relay will completely reset while the recloser is open following each instantaneous opening.

Recloser Operation	Percent of Travel Relay Travel
First Time-Delay Trip	(0.25/0.65) x (100) = +38.5%
Open for One Second	(1/6) x (100) = -16.7%
Second Time-Delay Trip	(0.25/0.65) x (100) = +38.5%

From this analysis, it appears that the relay will have a net travel of 60.3 percent of the total travel toward the contact-closed position.

From the foregoing, it is seen that the relay travel lacks approximately 40 percent (or 0.4 x 0.65 = 0.24 second) of that necessary for the relay to close its contacts and trip its breaker. On the basis of these figures, the IAC will be selective. A 0.15 to .2 second margin is generally considered desirable to guard against variations from published characteristics, errors in reading curves, etc. (the static overcurrent relay Type SFC overcomes some of these problems since the overtravel of such a relay is about 0.01 seconds and the reset time is 0.1 seconds or less).

If the automatic circuit reclosers are used at the substation as feeder breakers, it is necessary to select the proper size to meet the following conditions:

a. The interrupting capacity of the recloser should be greater than the maximum calculated fault current available on the bus.
b. The load-current rating (coil rating) of the recloser should be greater than the peak-load current of the circuit. It is recommended that the coil rating of the recloser be of sufficient size to allow for normal load growth and be relatively free from unnecessary tripping due to inrush current following a prolonged outage. The margin between

peak load on the circuit and the recloser rating is usually about 30 percent.

c. The minimum pickup current of the recloser is two times its coil rating. This determines its zone of protection as established by the minimum calculated fault current in the circuit. The minimum pickup rating should reach beyond the first-line recloser sectionalizing point, i.e., overlapping protection must be provided between the station recloser and the first-line recloser. If overlapping protection cannot be obtained when satisfying requirement (a), it will be necessary to relocate the first-line recloser to have it fall within the station's recloser protective zone.

Sectionalizers

The following basic coordination principles should be observed in the application of sectionalizers.

1. The minimum actuating current of sectionalizers should be 80% of the minimum trip of the source-side device. For electronically controlled units, the minimum actuating level is used directly. For hydraulically controlled units, the same series coil rating is used. The minimum actuating current is 1.6 times the sectionalizer coil rating to provide proper coordination with recloser minimum trip.

2. Sectionalizers not equipped with ground fault sensing should be coordinated with the phase pickup minimum trip level of the backup device. Setting the sectionalizer actuating level to coordinate with the backup device ground pickup level may cause erroneous lockout operations due to inrush current.

3. The sectionalizer should be set to lockout in one less operation than the backup device. This general rule need not apply in the case of several sectionalizers in series, where successive units may be set for one, two, or three operations less than the backup recloser.

4. The opening and reclosing times of the backup device must be coordinated with the sectionalizer's count retention time. The combined tripping (except for the first trip) and reclosing times of the backup must be shorter than the sectionalizer's memory time.

    If the backup operating time is longer than the sectionalizer's memory time, the sectionalizer will partially "forget" the number of backup tripping operations. This will result in the backup locking out for a fault beyond the sectionalizer and may require an extra backup tripping operation, and then both the backup device and the sectionalizer would be locked out.

5. Three-phase sectionalizers are limited to coordination with three-phase simultaneous opening backup devices. Nonsimultaneous tripping of backup devices could result in an attempted fault interruption by the sectionalizer, which is not designed for such operation.

Table 4-7. Sectionalizer Counts				
Sequence of Events				**Comment**
Step	**S1**	**S2**	**S3**	
1	0	0	0	Fault initiation
2	1	0	0	Breaker opens
3	1	0	0	Breaker closes
4	2	1	1	Breaker opens
5	2	1	1	Breaker closes
6	3	2	2	Breaker opens and sectionalizer number 1 opens
7	3	2	2	Breaker closes but inrush again produces pickup
8	3	3	3	Inrush produces a count and S2 and S3 try to open under load.

The sequence of events shown above indicates one of the problems sectionalizers suffer due to inrush. In this case (and there are others) the sectionalizers beyond the fault count incorrectly due to inrush. As can be seen, all sectionalizers could indeed open if the magnitudes of inrush were high enough.

One condition troublesome for sectionalizers without inrush restraint is shown in Figure 4-43. In this example a fault occurs on the lateral protected by sectionalizer #1. After the feeder breaker opens, this sectionalizer will count 1. The other sectionalizers will count 0, since they did not see fault current. If the fault is permanent the recloser will again close and open. This time, sectionalizer #1 will count 2 but sectionalizers #2 and #3 will count 1 (see Table 4-7) since inrush through them on reclose is of a magnitude similar to a fault current. This process continues until the breaker opens and sectionalizer #1 counts 3 and drops open, isolating the fault. The other sectionalizers, which have counted to 2, see another inrush during this successful reclose and try to drop open during a normal energized condition. Since some sectionalizers cannot interrupt load current this could also result in failure.

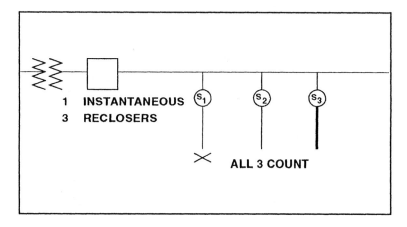

Figure 4-43. Sectionalizer Application

Example. The distribution feeder shown in Figure 4-44 illustrates a situation where a sectionalizer must coordinate with a 100 amp hydraulic recloser. The recloser is set for 1 fast and 3 time delay operations and the sectionalizer size must be selected.

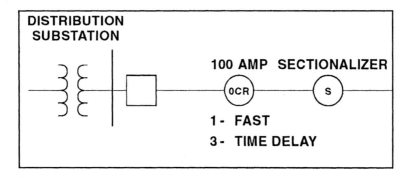

Figure 4-44

Minimum trip of recloser = 2 x rating
= 2 x 100 amps = 200 amps

Sectionalizer trip setting = 80% * 200 amps
= 160 amps

$$\text{Sectionalizer rating} = \frac{\text{minimum trip setting}}{1.6}$$ 4-10

$$= \frac{160}{1.6} = 100 \text{ amps}$$

Sectionalizer must be rated 100 amps or less.

QUESTIONS

1. Give two reasons why distribution system protection differs from subtransmission or transmission protection.

2. Explain why the fault current level drops off so much only a mile or so from the substation.

3. Expulsion fuses do not limit short circuit current. Explain.

165

4. The "general purpose" current limiting fuse has more low current interrupting capability than the "backup" current limiting fuse but is still not considered "full range". An expulsion fuse, on the other hand, has considerably less interrupting capability but is considered to be "full range". Explain.

5. Why must the arc voltage of the current limiting fuse be limited?

6. What is a "damage curve"? And when is it used?

7. Will a 10C "protecting fuse" coordinate with a 20C "protected fuse"?

8. What are the three required characteristics of a protective device?

9. Name two types of overcurrent relays.

10. Why should the relay curve be at least .3 seconds greater (for all current levels) than the downstream device?

11. What is "cold load pickup" and how is it considered in the setting of a relay?

12. Discuss the difference between a "recloser" and a breaker with a reclosing relay.

13. Explain why sectionalizers are easy to coordinate.

14. What is "fault selective feeder relaying" and how easy is it to obtain?

15. The minimum time for a breaker to clear a fault is about 6 cycles. Explain.

16. Impedance of a fault Zf is generally less than _____ ohms.

17. Why is "inrush" such a problem?

18. What is "notching"? How can the problem be minimized?

19. Pick a fuse for a 25 kVA, 1-phase transformer used on a 13.8 grounded wye system.

20. Pick a fuse size for a 1200 kVAR capacitor bank installed on a 34.5 kv grounded wye system.

5

SURGE ARRESTERS

CHARACTERISTICS OF LIGHTNING

In order to understand the effects of lightning, it is best to acquire some knowledge as to what lightning is, how it is caused, and where it is most likely to occur. Terms commonly used when describing lightning which will be discussed in this section are as follows:

- Stroke Leader
- Time Duration
- Current Magnitude
- Rate of Rise
- Multiple Strokes
- Polarity
- Isokeraunic Level

Stroke Leader

Under normal conditions, it is generally believed that clouds contain positive and negative charges that, being unlike, combine and neutralize each other resulting in substantially zero charge and hence zero voltage difference within the cloud.

One explanation of lightning is that when moist air is heated, it rises rapidly and when it gets to the higher altitudes it begins to cool. At very high altitudes (as high as 60,000 feet) precipitation particles are formed and begin to fall. This air going up and particles coming down (as fast as 100 mph) create the mechanism of transfer of charge where the cloud polarizes (see Figure 5-1).

When the voltage gradient between clouds or between clouds and earth reaches the limit for air, the air in the region of high stress ionizes and breaks down. The stroke leader, which is imperceptible to the eye, is a corona-like streamer that usually starts in the cloud as an electrical puncture. This in turn establishes the downward path of the lightning stroke between cloud and earth. The leader usually follows the direction

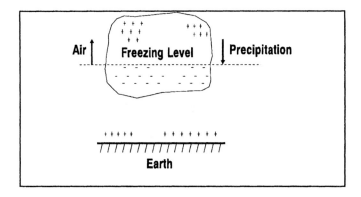

Figure 5-1. Separation of Charges Within Cloud

of the highest concentration of voltage gradient in successive steps. These zig-zag steps are approximately 50 yards at a time with 30 to 90 microsecond hesitations between steps as shown in Figure 5-2.

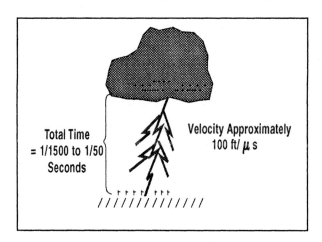

Figure 5-2. Propagation of Stroke Leader

As the leader approaches earth, negative ions progress downward along the leader path and positive ions begin propagating upward (upward strokes). When the upward and downward leaders meet as shown in Figure 5-3, a cloud to ground connection is established and the

169

charge energy from the cloud is released into the ground. This release of energy is the visible discharge we call lightning.

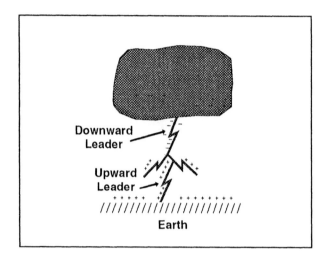

Figure 5-3. Return Strokes

Time Duration

The duration of a lightning stroke is usually less than a couple of hundred microseconds as shown in Figure 5-4. The industry accepted 8x20 current wave shown in Figure 5-4 is a reasonable approximation of a lightning surge.

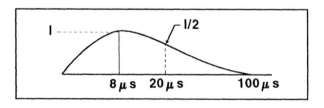

Figure 5-4. Typical Lightning Surge

Some lightning strokes have relatively high discharge currents over fairly short periods of time. These strokes usually produce shattering or explosive effects without much burning taking place. On the other hand, other lightning strokes last up to thousands of microseconds with fairly low discharge currents (< 1000 amps). This type of stroke, commonly called hot lightning, produces considerable burning, melting, fires, etc. Many lightning strokes are actually a combination of both.

The following data shows one of the statistical distributions of stroke duration reported to the industry.

Single Stroke Duration (µs)	Percent
> 20	96
> 40	57
> 60	14
> 80	5
Average time = 43 microseconds	

The energy in a lightning stroke is not nearly so high as some people imagine because the duration of the wave is so short. For example, a 43 microsecond wave lasts only .26 percent of the duration of a single cycle of 60 Hz current.

Current Magnitudes

The subject of current magnitudes is controversial and confusing. While most experts agree that stroke currents have been measured in excess of 200 kA, it is questionable to many as to how much of that current is seen by shielded equipment or discharged through an arrester.

Measurements of lightning currents over the years show that typical stroke currents fall into the following range:

Range of Stroke Currents
5% exceeded 90,000 amperes
10% exceeded 75,000 amperes
20% exceeded 60,000 amperes
50% exceeded 45,000 amperes
70% exceeded 30,000 amperes

Rate of Rise

It is interesting to note that, while the industry tests on an 8x20 µs wave, this wave shape is not at all substantiated by field data. Times to crest shown in Figure 5-5 are much more representative than the 8x20 µs wave. All the investigators we have researched reported rates of rise higher than 10 kA/µs for over 50 percent of stroke currents. Rates of 65 kA/µs for many stroke currents were reported by several of the investigators.

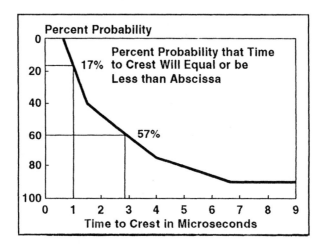

Figure 5-5. Time to Crest vs. Probability

Multiple Strokes

Over half of all lightning strokes are multiple and have from two to as many as forty-two strokes. Multiple strokes are caused by quickly recharging the cloud area. After the stroke first occurs, some of the electric charges in other parts of the same, or adjacent, clouds move in to replenish the discharged area. This replenishment occurs before the gaseous path of the first stroke has dissipated and consequently makes the path of each succeeding stroke approximately the same. Some typical values for multiple strokes are as follows:

- 50% of direct strokes had at least 3 components
- 24% of direct strokes had at least 4 components

- 15% of direct strokes had at least 6 components
- Average duration for multiple strokes was approximately 1/10 second
- Maximum duration for multiple strokes was approximately 1-1/2 seconds

Polarity

Polarity refers to the clouds and the earth's relationship in that the earth's charge is positive and the cloud's charge is negative in 90% of the recorded measurements.

Isokeraunic Level

The isokeraunic map shown in Figure 5-6 is used to indicate relative frequency of lightning on a geographical basis. The number of thunderstorm days in any particular location is known as the isokeraunic level. For example, Miami, Florida would expect to see between 70 and 80 thunderstorms each year. Many believe that this is only part of the story because there is indication that higher isokeraunic levels also result in more strokes per thunderstorm day.

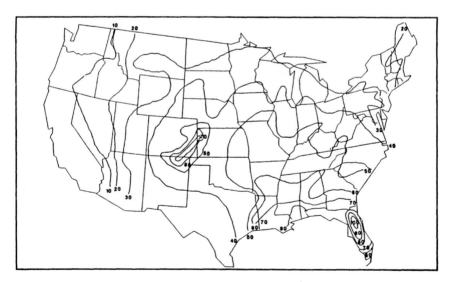

Figure 5-6. Isokeraunic Map

The United States takes a pretty bad beating with an average for the entire country of about 40 storms per year. The lightning season of much of the northern U.S. extends from early May to October, but prevails increasingly longer for regions farther south. In the Gulf states, lightning storms may occur from early March to December.

SYSTEM OVERVOLTAGE

There are several sources of overvoltages which must be considered when applying arresters. Overvoltages caused by lightning, neutral displacement during line-to-ground faults and current limiting fuse operation will be discussed later in the text. Several other sources of overvoltages which will be discussed in this section are as follows:

- Ferroresonance
- Capacitor Switching
- Current Chopping
- Accidental Contacts with Higher Voltage Systems.

Ferroresonance

In three-phase circuits, single-phase switching, fuse blowing, or a broken conductor can result in overvoltages when ferroresonance occurs between the magnetizing impedance of the transformer and the system capacitance of the isolated phase or phases.

A myriad of practical circuit situations can occur which may result in ferroresonance phenomena. Basically, the necessary conditions can arise when one or two open phases result in capacitance being energized in series with the nonlinear magnetizing impedance of a transformer, as in Figure 5-7, where the switches could be fuse cutouts at a cable riser pole, the capacitance could be that of a length of cable connecting to the ungrounded winding of a pad-mounted transformer.

Ferroresonance cannot be entirely avoided. Conditions that are likely to produce ferroresonance are as follows:

- Small transformer rating; the smaller the rating the greater the susceptibility. Bank ratings larger than 300 kVA are rarely susceptible.

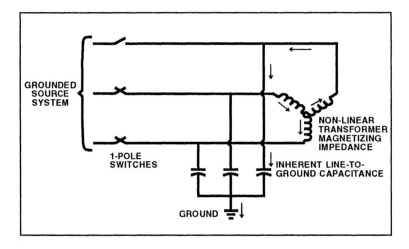

Figure 5-7. Single-Phase Switching in 3-Phase Circuit

- No load; a load as small as 4% will result in near immunity.
- Any 3-phase connection is susceptible. A single-phase transformer connected phase-to-phase on a grounded primary system is susceptible.
- Primary supply by underground cables; shielding increases the capacitance and susceptibility. Overhead primary cables generally provide immunity unless voltage is higher than 15 kV. At 34.5 kV FR is definitely a possibility with overhead supply since internal capacitance of transformer is sufficient for resonance.
- Primary voltage above 5 kV. Voltages below 5 kV provide substantial immunity. Above 15 kV, FR is quite likely. Opinions differ on susceptibility in the 5 to 15 kV range.
- Secondary capacitor bank with floating neutral, even on a grounded Y-grounded Y connection if one phase of primary is de-energized can energize magnetizing reactance of the de-energized phase through the capacitance and thus cause resonance.

At the present time, the most practical method of avoiding FR is by installation of wye-wye connected transformers with transformer primary and secondary neutrals grounded and connected to the system primary neutral. The primary grounded wye neutral "shorts out" the series connection of the transformer reactance and cable capacitance, thereby

175

preventing the establishment of a resonant circuit. Other mitigating techniques are as follows:

- Single-phase transformers should be connected line-to-neutral.
- FR can also be minimized by the installation of 3-phase switching and protective devices so that single phasing cannot occur. This may not be possible in many cases and it may not be completely effective, but it is the single best preventive.
- FR can be prevented if cables and transformers are never switched together. To accomplish this, the transformer switches need to be located at the transformer terminals rather than at the riser pole. If switches are also required at the riser pole, interlocking is desirable to ensure that upon energizing, all riser pole phases are first closed, and then all transformer primary switches are closed. Upon de-energizing, all phases must first be opened at the transformer before opening at the riser pole.
- The requirement for never switching primary supply cable and transformers together also applies to fuses and other protective devices. This requires that fuses, reclosers or sectionalizers at riser poles and on the distribution feeder are coordinated to hold-in on a transformer fault so the transformer primary protection will clear first. Of course, a primary cable fault would blow a fuse at the riser pole first, but such a fault is most likely to short out the capacitance of the faulted cable section connected to the transformer, and thus prevent resonance.
- If a susceptible connection must be used, and if primary cable runs are long and must be switched with the transformer, and if 3-phase switching and protection is not possible, then arrange the system to have all switching done with at least 5% loading on the transformer.
- Secondary capacitor banks should be connected with grounded neutral.

Capacitor Switching

Capacitor bank switching may cause an overvoltage upon either energization or de-energization. For example, consider the following grounded neutral bank energization (Figure 5-8).

If the initial conditions (pre-closing) are such that the capacitor bank has no charge (no voltage) and the system voltage at contact closure is at a maximum, the voltage will overshoot as shown in Figure 5-9.

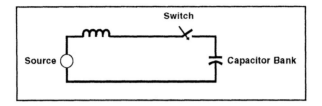

Figure 5-8. Capacitor Energization

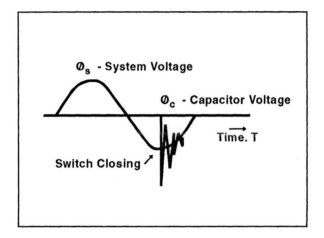

Figure 5-9. Capacitor Voltage During Energization

De-energization of a capacitor bank is even a greater concern. Going to the system representation in Figure 5-10, but now showing the switch opening, we create conditions sometimes referred to as "voltage magnification."

Assume that R and X_L are very small compared to the capacitive reactance so that the capacitor steady-state voltage is essentially the same as the source voltage. If the switch is assumed to have opened at some time shortly before time 0, then current interruption will take place at a "normal" zero such as at time "a" in Figure 5-11.

177

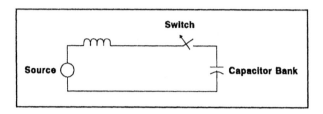

Figure 5-10. Capacitor De-energization

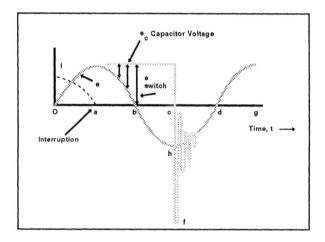

*Figure 5-11. Overvoltages Due to
De-energization*

By our assumptions, steady-state 60 Hz current leads the source voltage by 90 degrees as shown, so that the source (and capacitor) voltage reaches its maximum value at time "a". The result of interruption is that the capacitor voltage remains at peak value because of the charge which is "trapped" on it. However, the source voltage continues its normal 60 Hz variation, and the voltage which gradually appears across the switch is the difference between the fixed capacitor voltage on the one side and the source voltage on the other. As shown, the switch voltage reaches a maximum of twice normal value at time "c", one-half cycle following interruption.

If the switch can withstand twice normal voltage at this time, successful interruption has been achieved. Because of discharge resistors, normally built into the capacitors, the capacitor voltage will drain off, ultimately to disappear.

However, if the switch does not achieve adequate dielectric recovery, the arc may reignite or "restrike" between the contacts sometime during the period of "a" to "c", which will re-energize the capacitor. Maximum transient voltages result if a restrike takes place at maximum switch voltage, time "c". When current is re-established at this time, the capacitor voltage which is at plus 1.0 attempts to rejoin the system voltage at minus 1.0 or "h". It must travel 2.0 to reach value "h", and thus can overshoot point "h" by 2.0. Then the resulting voltage at "f" is 3.0 times normal.

Since the capacitor current also undergoes a natural-frequency oscillation, it is theoretically possible that a "natural-frequency" current zero may occur just after time "c". A second interruption here could leave a trapped charge on the capacitor with voltage "f" of negative 3.0 per unit. As the system voltage again swings to plus 1.0, a maximum switch voltage of 4.0 could result, and a restrike at time "g" would give 4.0 + 1.0 = 5.0 times normal voltage, etc. However, compounding of this nature is rarely, if ever, found in practice. Modern switches generally do not restrike or restrike more than once during clearing. Voltages approaching 3 times normal will occur only if restriking occurs at the worst possible time. Voltages in the order of 2.5 times normal are more typical of field measurements.

Current Chopping

Most fault-current interrupting devices, such as expulsion fuses, reclosers, circuit breakers, etc., accomplish their arc extinction by waiting for a 60 Hz current zero. Transients produced in this manner are usually twice normal or less. It is possible under some conditions such as current limiting fuse operation or breaker interruption of low currents where a current interruption occurs prior to the normal current zero. This so-called "current chop" can cause exceptionally high voltages depending on the rate of current interruption, the amount of current chopped and the system configuration.

Let us analyze abrupt current chopping by assuming that the current is forced instantaneously to zero from some finite value. If this current is flowing in an inductance, it cannot change instantaneously, and it therefore follows that practically there must be capacitance and/or resistance associated with the inductance if arc voltage is neglected.

Consider the circuit of Figure 5-12 where resistance is ignored and assume that the capacitive reactance is so much larger than the inductive reactance that its normal current is negligibly small compared to i (i.e., $\omega_n = 1/\sqrt{LC}$ is very large).

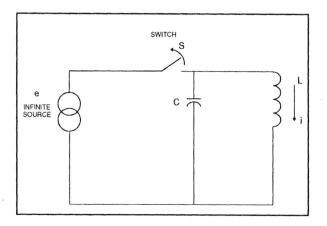

Figure 5-12. Circuit Illustrating Current Chopping

When the switch current suddenly changes from a value i to zero, i continues to flow instantaneously in L; therefore, it must also flow in C. A natural frequency oscillation ensues in L and C. The maximum natural frequency voltage appearing across L (and C) is

$$e = \sqrt{\frac{L}{C}}\, i \qquad\qquad 5\text{-}1$$

Thus, the voltage is proportional to the magnitude of current chopped and to the surge impedance of the circuit being switched.

This equation can be manipulated as follows to express the transient voltage in a different way.

$$e = \sqrt{\frac{L}{C}}\ i = \sqrt{\frac{\omega L}{\omega C}}\ i = \sqrt{X_L * X_C}\ i$$

5-2

$$= X_L \sqrt{\frac{X_C}{X_L}}\ i$$

However, $\sqrt{\dfrac{X_C}{X_L}}$ is the per unit natural frequency, ω_n/ω.

Thus,

$$e = \frac{\omega_n}{\omega}\ X_L\ i$$

5-3

If $X_L i$ is the normal voltage, or some measure of the normal voltage across the inductance, then the transient voltage, e, is many times normal by the ratio ω_n/ω.

Theoretically, then, chopping can produce very high voltages. In practice, however, L is often the nonlinear magnetizing impedance of a transformer. The magnetic characteristics of modern transformers coupled with typical switch performance do not usually give rise to voltages more than 2 times normal.

Accidental Contact with Higher Voltage Systems

Often overhead primary distribution circuits are built underneath higher voltage circuits on the same pole. Broken high-voltage conductors can fall upon the lower voltage circuit primary possibly causing the lower rated arresters to fail along the entire line or other major equipment damage.

SILICON CARBIDE VS. MOV ARRESTERS

The utility industry, like other industries has seen changes in arrester design over the years. While several older designs can still be found on a few distribution systems, the vast majority of arresters now in use are either:

• Gapped Silicon Carbide, or
• MOVs (Metal Oxide Varistors).

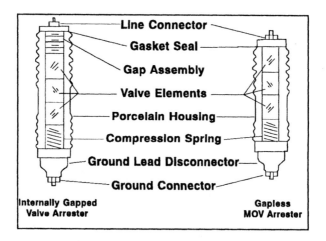

Line Connector
Gasket Seal
Gap Assembly
Valve Elements
Porcelain Housing
Compression Spring
Ground Lead Disconnector
Ground Connector

Internally Gapped Valve Arrester

Gapless MOV Arrester

Figure 5-13. Silicon Carbide and MOV Arrester

Most of the arresters found on the distribution system are the older type gapped silicon carbide. The introduction of the metal oxide arrester in the late 70's is one of the most significant advances in the utility industry especially since its acceptance by the industry became almost universal by the early 1980's. At this time, manufacturers have even gone completely out of the business of making intermediate and station class silicon carbide arresters and very few are making the distribution type.

A silicon carbide surge arrester has silicon carbide valve elements that are protected from continuous power-frequency voltage by a series gap which acts as the insulator during normal voltage conditions and interrupts the power-frequency current that follows any transient current discharged by the arrester. It does this by not restriking on subsequent half cycles of power-frequency voltage after the first follow-current zero has occurred. Voltage and current zeroes occur simultaneously, permitting the gap to clear the circuit established through the arrester.

In a metal-oxide-varistor arrester, the metal-oxide disk insulates the arrester electrically from ground. The disk is composed of a variety of materials in varying concentrations which determine the electrical characteristics of the varistor. Highly conductive particles (usually zinc oxide) are suspended in a true semiconductor with characteristics close to that of a back-to-back zener diode.

The processing of metal-oxide-varistor disks is extremely critical. Purity of the materials and their uniform dispersion throughout the disk must be carefully monitored. To demonstrate how critical monitoring is, an experiment was conducted in which the concentration of one of the materials in a disk was increased by 50 parts per million beyond its specified parts-per-million concentration. The varistor that resulted from this increase demonstrated a 15% improvement in protective characteristics, but the overall life expectancy of the varistor was decreased by 90%.

A metal-oxide disk goes into conduction sharply at a precise voltage level and ceases to conduct when voltage drops below this level. A series gap is not required to insulate the arrester from ground or to interrupt power follow current (which does not exist as long as applied power-frequency voltage is below conduction voltage level).

The primary difference between silicon carbide and MOV arresters is that the MOV valve blocks are so nonlinear that no or at least very little power current is drawn at normal line-to-ground voltage. The MOV arrester consequently does not require a series gap. The MOV simply eases in and out of conduction. A comparison of the nonlinear characteristics shown in Figure 5-14 dramatically illustrates the extreme nonlinearity of the MOV.

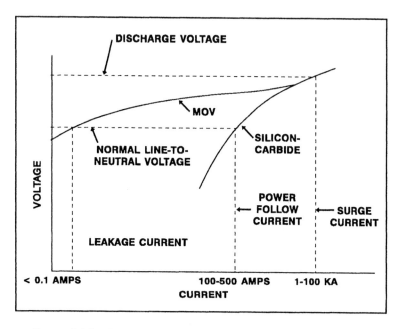

Figure 5-14. Comparison of the Nonlinear Characteristics of
MOV and Conventional Silicon Carbide Valve Blocks

CLASSES OF ARRESTERS

There are 3 classes of arresters: distribution, intermediate and station.
For the most part the major difference in these types of arresters is the
size of the block. A larger block reduces IR discharge voltage and greatly
increases energy capability and thus reliability. All three classes of
arresters are available for distribution system as shown below.
Distribution class arresters are as a rule utilized out on the distribution
feeder whereas intermediate and station class arresters are used in the
substation.

Table 5-1. Voltage Ratings in Kilovolts		
Distribution Arresters	**Intermediate Arresters**	**Station Arresters**
1		
3	3	3
6	6	6
9	9	9
10		
12	12	12
15	15	15
18		
21	21	21
	24	24
25		
27		
30	30	30
	36	36
	39	39
	48	48
	60	60
	72	72
	90	90
	96	96
	108	108
	120	120

ARRESTER SELECTION

Choosing an arrester rating for a distribution system is based on the system's line-to-ground voltage and the way it is grounded. The limiting condition for an arrester does not usually have anything to do with the magnitude of the surges, (switching or lightning) that it might see. This is in contrast to the selection of arresters for transmission. In distribution, rating of the arrester is based on the maximum steady state line-to-ground voltage the arrester might see. This limiting condition is normally caused when there is a line-to-ground fault on one of the other phases.

According to ANSI Standard C62.22, "Guide for the Application of Metal-Oxide Surge Arresters for Alternating-Current Systems", proper application of arresters on distribution systems requires knowledge of 1) the maximum normal operating voltage of the power system, and 2) the magnitude and duration of temporary overvoltages (TOV) during abnormal operating conditions. This information must be compared to the arrester MCOV rating and to the arrester TOV capability.

MCOV

The term MCOV or "maximum continuous operating voltage" sounds simple enough but has been difficult for many utilities to determine. On a distribution system where the voltage is always changing due to varying load demands, and where the voltage on one part of the system may be somewhat different to other parts (e.g., near the substation and at the end of the feeder), it is sometimes impossible to define only one MCOV.

The MCOV of the arrester is, however, somewhat easier to define since it is approximately 84% of the arrester duty cycle rating as shown in Table 5-2. What this means is that a 10 kV duty cycle rated arrester, typically used for a 13.2 kV system could be operated continuously with a maximum continuous line-to-ground voltage of 8.4 kV or less.

Table 5-2. Arrester	
Arrester Rating	**MCOV**
3	2.55
6	5.10
9	7.65
10	8.40
12	10.2
15	12.7
18	15.3
21	17.0
24	19.5
27	22.0
30	24.4

Table 5-3. Commonly Applied Voltage Ratings of Metal-Oxide Arresters on Distribution Systems*				
System Voltage (Kilovolts rms)		Commonly Applied Arrester Voltage Ratings - kV RMS Duty Cycle Voltage Ratings (MCOV)†††		
Nominal Voltage	Maximum Voltage Range B**	Four-Wire Multigrounded Neutral Wye	Three-Wire Low Impedance† Grounded***	Three-Wire High Impedance† Grounded
2400	2540			3 (2.55)
4160Y/2400	4400Y/2540	3 (2.55)	6 (5.1)	6 (5.1)
4260	4400			6 (5.1)
4800	5080			6 (5.1)
6900	7260			9 (7.65)
8320Y/4800	8800Y/5080	6 (5.1)	9 (7.65)	
12000Y/6930	12700Y/7330	9 (7.65)	12 (10.2)††	
12470Y/7200	13200Y/7620	9 (7.65) OR 10 (8.4)	15 (12.7)††	
13200Y/7620	13970Y/8070	10 (8.4)	15 (12.7)††	
13800Y/7970	14605Y/8430	12 (10.1)	15 (12.7)††	
13800	14520			18 (15.3)
20780Y/12000	22000Y/12700	15 (12.7)	21 (17.0)††	
22860Y/13200	24200Y/13870	18 (15.3)	24 (19.5)††	
23000	24340			30 (24.4)
24940Y/14400	26400Y/15240	18 (15.3)	27 (22.0)††	
27600Y/15930	29255Y/16890	21 (17.0)	30 (24.4)††	
34500Y/19920	36510Y/21080	27 (22.0)	36 (29.0)††	

*   Spacer cable circuits have not been included - there has been insufficient experience with the application of metal-oxide arresters on spacer cable circuits to include them in this table - refer to [30] for information on spacer cable circuit overvoltages.

**   See ANSI C84.1-1989.

***   Line-to-ground fault duration not to exceed 30 minutes. For longer durations consult manufacturers' temporary overvoltage capability.

†   Low impedance circuits are typically 3-wire, unigrounded at the source. High impedance circuits are generally ungrounded (i.e., delta). Additional information regarding system grounding is contained in ANSI C62.92, Part 1.

††   Individual case studies may show lower voltage ratings may be used.

†††   For each Duty Cycle rating the Maximum Continuous Operating Voltage (MCOV) is also listed.

Table 5-3, from ANSI C62.22, shows the commonly applied voltage ratings of metal-oxide arresters for distribution systems. All these duty cycle ratings are the same as the rating for the older gapped silicon carbide arresters except at the 13.8 kV level. Typically, a 13.8 kV, 4-wire, multigrounded system have used 10 kV gapped arresters. Today, most of these same utilities are still using 10 kV MOVs. Some utilities, however, have recognized that the 10 kV arrester is very marginal and should possibly be replaced by a 12 kV rating to be on the more conservative side.

TOV

How much voltage shift which will occur is a function of the type of system grounding. For example, on a delta system, a line-to-ground fault will cause a full offset, i.e., the line-to-ground voltage will become the line-to-line voltage. Figure 5-15, shown below, illustrates this condition. As can be seen, when a phase has a fault there is no current since the transformer is delta connected. In a sense, ground at this point is A phase. The arresters connected from B and C phase to ground will now in effect be connected B to A phase and C to A phase or line-to-line. This of course means that the voltage across these arresters will increase to 1.73 per unit. This is the voltage upon which the arrester rating is based whether they be gapped silicon carbide or MOV's.

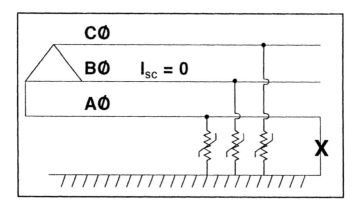

Figure 5-15. Line-to-Ground Fault on a Delta System

Most distribution systems are classified as 4-wire multigrounded systems. The fourth wire is, of course, the neutral wire which is grounded periodically at the pole. Figure 5-16 illustrates this type of system under a single line-to-ground fault condition. As can be seen, for this solidly grounded system considerable fault current will flow. If grounding was perfect, there would be no voltage difference between the substation ground and the point of the fault. If this were the case, the voltage at the point of the fault would remain at 0 potential and arresters connected from the other two phases would see no change in voltage. The ground, however, is not perfect and some rise does occur. For this type of system, the rise associated with a single line-to-ground fault is considered to be 20% (or 25% if regulation is factored in). Consequently, the arresters on B and C phases would see approximately 1.25 per unit across their terminals for this condition.

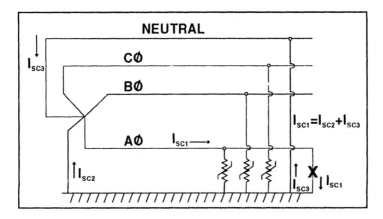

Figure 5-16. Line-to-Ground Fault on a Grounded Wye System

Based upon some of the concepts just discussed as well as many years of successful operating experience, a proposal for the selection of distribution lightning arresters was made some years ago by a working group in IEEE. This group proposed that the ratings of gapped type surge arresters selected for open-wire, multigrounded neutral systems be equal to or greater than the nominal line-to-neutral voltage multiplied by the product of the regulation factor 1.05 and the voltage rise factor 1.2. This is equivalent to 1.25 times nominal line-to-neutral system voltage. For an MOV type arrester this voltage is compared to the TOV rating of the MOV. Because the MOV arrester is more sensitive to poor grounding, poor regulation, and the reduced saturation sometimes found in new transformers, it is generally recommended that a 1.35 factor be considered for MOVs.

A summary of this and other recommendations is as follows:

- **Open-Wire Multigrounded System**
  Rating = Nominal L-G voltage x 1.25 (Gapped)
  Rating = Nominal L-G voltage x 1.35 (MOV)
- **Spacer Cable Systems**
  Rating = Nominal L-G voltage x 1.5
- Unigrounded Rating = Nominal L-G voltage x 1.4

The difficulty now is to determine the duration of the TOV. If the assumption is made that the maximum duration of the fault is 300 seconds, then the maximum temporary system overvoltage could be 1.095 per unit of duty cycle rating (from Figure 5-17) or 1.3 per unit of MCOV, since duty cycle rating is approximately 19% higher than MCOV.

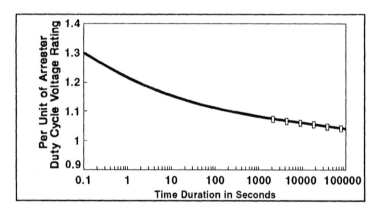

Figure 5-17. Example Curve - Consult Manufacturer for
Exact Curve

An example of calculations to determine the arrester rating is as follows:

Table 5-4						
System Voltage kV	System Grounding	Multiplication Factor	Max I-g Voltage kV	Min. Required MCOV*	Arrester Duty Cycle Rating	Actual Arrester MCOV
12.47	Multigrounded	1.35	9.72	7.5	9	7.65
13.2	Multigrounded	1.35	10.3	7.9	10	8.4
13.8	Multigrounded	1.35	10.76	8.3	10	8.4
23.	Multigrounded	1.35	17.9	13.8	18	15.3
34.5	Multigrounded	1.35	26.9	20.7	27	22.0
34.5	Delta	1.82	36.25	27.9	36	29.0
34.5	Spacer	1.5	29.9	23.	30	24.4
34.5	Uniground	1.4	27.8	21.4	27	22.0
*Min MCOV = Max I-g voltage/1.3						

Selection

As can be seen, when comparing Table 5-3 and Table 5-4, the arrester ratings for four wire multigrounded systems tend to be determined by the MCOV while the less effectively grounded systems tend to be determined using TOV. A summary of recommended arrester ratings for various system voltages and grounding practices and faults lasting less than 300 seconds is shown in Table 5-5.

Table 5-5. Arrester Rating (Duty Cycle)*				
System Voltage kV	4-Wire Multigrounded	Delta	Unigrounded	Spacer
12.47	9	12	10	10
13.2	10	15	10	12
13.8	12	15	12	12
23	18	24	18	21
34.5	27	36	27	30
*Rating based on fault lasting <300 seconds.				

Considerations in the Applications of MOVs

Selection of an MOV arrester is primarily based upon the maximum continuous operating voltage (MCOV) that is applied across the arrester in service (line-to-ground). For arresters on effectively grounded systems, this is normally the maximum line to ground voltage (e.g., on a 13.8 kV system the maximum steady state line-to-ground voltage is calculated as $1.05 * 13.8 / \sqrt{3}$ or 8.37 kV). For an ungrounded or impedance-grounded system, the MCOV should be at least 90 percent of the maximum phase-to-phase voltage.

Other considerations in the application of MOV's are temporary conditions on the distribution system that raise the voltage such as a line-to-ground fault. MOV's have temporary overvoltage curves similar to the one shown below which must be considered for the following conditions:

Voltage Regulation. Voltage standards call for a voltage at the meter no higher than 5% over nominal. This standard does not however, limit the voltage fluctuation out on the feeder. For example, an Electric Power Research Institute study showed that the substation voltage can be as high as 17% over nominal (the average was 7%). With average substation voltages, voltages out on the feeder are generally not more than 5% over nominal. It is suspected that capacitor operation at light loads, or improper settings of voltage regulators are resulting in system voltages 10% above nominal or even higher without the knowledge of the utility. The primary concern regarding these quasi-steady-state overvoltages is that long term stability of metal oxide valve elements is normally demonstrated at MCOV rather than at these higher voltages. A utility suspecting such overvoltages may do well to investigate going to the next higher rated arresters.

Ferroresonance. Ferroresonance has long been a concern of the distribution engineer. Higher voltage levels, longer lines, and underground cables have further increased this concern. Ferroresonance, when it occurs, does not always result in component failure. This is due in part to the fact that the overvoltages can be fairly low and the condition may not last more than a few seconds, e.g., closing or opening a switch. Most ferroresonant overvoltages are in the range of 1.5 to 2.0 P.U. and as noted earlier, the power frequency sparkover of gapped silicon carbide arresters is usually 2.0 to 2.4 P.U. The effect of ferroresonance may then be negligible in many cases because the sparkover voltage of the arrester is rarely, if ever, exceeded. On the other hand, since a metal oxide arrester has no gap and conducts current that is a function of applied voltage, it will make an unsuccessful effort to reduce the ferroresonance overvoltage. The current conducted by a metal oxide distribution arrester at 2.0 P.U. may be up to a few hundred amperes, depending on the impedance of the system, and the arrester may fail unless steps are taken to limit the possible duration of the overvoltages.

Cogeneration. Overvoltages may occur in many circuits primarily because of problems associated with relay application and operation. For example, cogenerators are often interconnected to the primary distribution system by means of a delta-connected transformer. This connection utilizes only 3 wires of the standard four-wire system. During a feeder ground fault, the feeder breaker will open to separate the faulted section from the system, but because of the delta connection, the overcurrent protection of the cogenerator will not operate. After some delay, the underfrequency or undervoltage relay of the cogenerator will operate to disconnect it from the faulted section. During the period of time between feeder breaker operation and cogenerator separation, the four-wire faulted section operates as a three-wire system and the line-to-ground voltages on the unfaulted phases may reach 1.73 P.U. This overvoltage may have no effect on gapped arresters, but the metal oxide arresters on this feeder will conduct current on the overvoltage. If the overvoltage is high, the metal oxide arresters will fail unless the duration of the overvoltage is short.

Line-to-Ground Faults. Selection of rating for gapped silicon carbide distribution arresters is based on experience and on calculated values of overvoltage on the unfaulted phases of a three-phase circuit during a fault to ground on one phase. The most commonly used application rule for open-wire multigrounded neutral systems is that arrester rating be "equal to or greater than the product of the nominal [line-to-neutral voltage] x [the Range A factor 1.05] x [1.2] which is the maximum voltage rise on the unfaulted phases of a loaded circuit. This is equivalent to 1.25 times nominal line-to-neutral system voltage. The rule cited above is very conservative for silicon carbide arresters because the maximum calculated voltage is equal to arrester rating, and arrester sparkover is well in excess of rating; therefore, another transient must be superimposed to cause sparkover during the short time that the overvoltage exists. Furthermore, silicon carbide arresters will generally tolerate at least a few operations at overvoltages as high as 1.2 times rating (1.5 x nominal line-to-neutral voltage). For these reasons, the application rule given above has been satisfactory for silicon carbide arresters even though many four-wire distribution systems are not effectively grounded. The same rule is not as conservative for metal oxide arresters unless it is known that the system is truly grounded. Some metal oxide arrester failures are being experienced with the cause of failure believed to be overvoltages in excess of arrester temporary overvoltage capability. In order to obtain an accurate estimate of system overvoltage for comparison with arrester temporary overvoltage capability, it is necessary to take into account factors such as ground wire size, ground rod spacing, fault resistance, earth resistivity and system impedance. Some utilities are using a 1.35 factor instead of the usual 1.25 to accommodate for this.

INSULATION COORDINATION

Margins for Overhead Equipment

From previous sections, we know how an arrester is rated, but how is it applied? It is important to note that application of arresters for transmission and distribution is different. In transmission, lightning is of secondary concern in surge arrester application. Primary concern is switching surges. On a distribution circuit, however, the relative low-voltage and short lines tend to make switching surges minimal and, consequently, lightning is of primary importance.

Reflection of this fact can be seen in typical characteristics published for distribution-class arresters as shown in Tables 5-6A and 5-6B.

As can be seen, protective characteristics are shown for front-of-wave sparkover and IR discharges, but not for switching surge waves (as shown for higher rated transmission arresters).

The two protective characteristics normally used for "insulator coordination" are:

Front-of-Wave Sparkover. This is the first thing that happens to the gapped arrester -- it sparks over. It is compared to the fast front equipment insulation characteristics such as the chopped wave insulation

Table 5-6A. Distribution Arrester Characteristics per Handbook - Silicon Carbide					
Arrester Rating (kV RMS)	Maximum ANSI Front-of-Wave Sparkover (kV Crest)		Maximum Discharge Voltage (kV Crest) at Indicated 8 x 20 Microsecond Impulse Current		
	With Disconnector	Externally Gapped	5000 Amperes	10,000 Amperes	20,000 Amperes
3	14.5	31	11	12	13.5
6	28	51	22	24	27
9	39	64	33	36	40
10	43	64	33	36	40
12	54	77	44	48	54
15	63	91	50	54	61
18	75	105	61	66	74
21	89	...	72	78	88
27	98	...	87	96	107

Table 5-6B. Distribution Arrester Characteristics per Handbook - MOV (Heavy Duty)					
Arrester Rating	MCOV	Front-of-Wave Protective Level*	Maximum Discharge Voltage 8/20 µs Current Wave		
kV rms	kV rms	kV Crest	5 kA	10 kA	20 kA
3	2.55	10.7	9.2	10.0	11.3
6	5.10	21.4	18.4	20.0	22.5
9	7.65	32.1	27.5	30.0	33.8
10	8.40	35.3	30.3	33.0	37.2
12	10.2	42.8	36.7	40.0	45.0
15	12.7	53.5	45.9	50.0	56.3
18	15.3	64.2	55.1	60.0	67.6
21	17.0	74.9	64.3	70.0	78.8
24	19.5	84.3	72.3	78.8	88.7
27	22.0	95.2	81.7	89.0	100.2
30	24.4	105.9	90.9	99.0	111.5
36	30.4	124.8	107.0	116.6	131.3

*Based on a 10 kA current impulse that results in a discharge voltage cresting in 0.5 µs

level of the transformer. An MOV has no gap but does have an equivalent sparkover as shown in Table 5-6B.

IR Discharge at 10 kA. After the arrester sparks over the gap, the lightning current discharges through the block material. Standards recommend that a 10 kA discharge level be used for coordination purposes. Many utilities, however, are using a 20 kA discharge level to gain some additional margin. (Discharge characteristics across a MOV are very similar so margin calculation is virtually identical.)

Distribution equipments are normally defined as being in a voltage class such as 15 kV, 25 kV, etc. Most utility equipment is operated in the 15 kV class. A distribution transformer in the 15 kV class is defined by the following insulation characteristics:

*   60 Hz, one minute withstand = 34 kV
*   Chopped Wave (short-time) = 110 kV at 1.8 μs
*   BIL = 95 kV

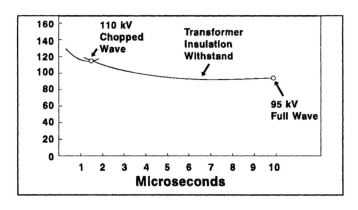

Figure 5-18. Characteristics of Transformer Insulation

Assuming a 12,470 volt, 4-wire system (7200 volts L-G), we would select the arrester rating based on the rules developed in the previous section, i.e., a 9 kV arrester (gapped or MOV).

We can see from Table 5-6 that a 9 kV gapped arrester has a sparkover of 39 kV and an IR discharge at 10 kA of 36 kV. This could be plotted with the transformer characteristics as follows:

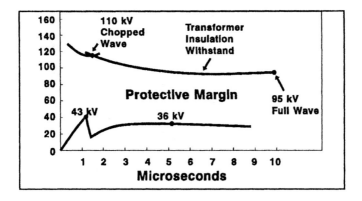

Figure 5-19. Insulation Coordination

Standards recommend 20% margins calculated by the formula:

$$\text{Margin} = \frac{\text{Insulation Withstand} - \text{Protective Level}}{\text{Protective Level}} \times 100 \qquad 5\text{-}4$$

Two margins are calculated, one for the chopped wave and one for the full wave (BIL) of the transformer. These calculations are performed as follows:

$$\% \text{ Margin} = \frac{110 - 39}{39} \times 100\% = 182\% \text{ (Chopped Wave)}$$

$$5\text{-}5$$

$$\% \text{ Margin} = \frac{95 - 36}{36} \times 100\% = 164\% \text{ (BIL)}$$

As can be seen, these margins (182% and 164%) are greatly in excess of the recommended 20% and consequently show good protection practice. If we were using an MOV we would simply use the equivalent sparkover or compare only the IR discharge and the BIL since this is the lesser margin. The margins would be similar.

Margins for Underground Equipment

If the system is underground, we must be more concerned with the phenomena of traveling waves and the consequent doubling of voltage surges at an open point. For example, a typical underground residential design is shown in Figure 5-20.

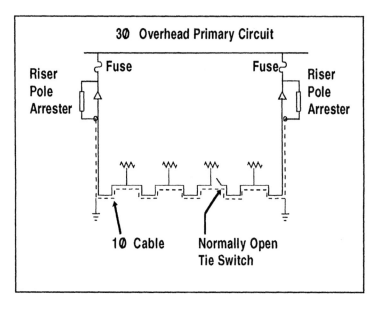

Figure 5-20. Underground Lateral

A surge entering this cable will travel to the open point where its voltage will double, as shown in Figure 5-20 and start on its way back.

This reflected wave plus the incoming waves impose approximately twice the normal voltage on the entire cable and all of the equipment connected to it (see Figure 5-21). For example, if we had an arrester with a 36 kV IR discharge level (we are only considering BIL margin), we would now expect to see 72 kV imposed across the insulation of this equipment. The new margin would then be calculated as follows:

$$\% \text{ Margin} = \frac{95 - 72}{72} \times 100 = 32\% \qquad\qquad 5\text{-}6$$

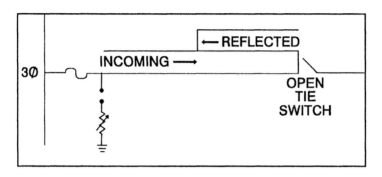

Figure 5-21. Reflected Surge Voltage at Open Point

Since 32% is greater than the recommended 20% no problem would normally be anticipated. However, it is well known to many utilities that impulse failures at this voltage do occur.

Similar margins can be calculated for other classes of voltages as shown in Table 5-7.

Table 5-7					
Distribution Voltage 4-Wire Multiground	BIL	Arrester Rating	IR at 10 kA	Overhead Margin %	Underground Margin %
12470	95	9	36	164	32
24940	125	18	66	89	-5
34500	150	27	96	56	-22

As can be seen, the only underground systems that can be protected using a standard distribution class arrester are those in the 15 kV class and below. At the 34.5 kV and 25 kV levels the margins are actually negative. The general recommendations at these high voltage levels are:

- Use a better riser pole arrester (intermediate or station class).
- Put an arrester at the open point.

199

Use of an intermediate class arrester at the riser pole can create slightly more margin but is usually not sufficient to maintain the recommended 20% level. Open tie arresters on the other hand do not totally prevent doubling but are still quite effective in increasing margin.

On 25 kV underground systems, many utilities still use only riser pole protection but put in an intermediate class arrester (IR ≈ 55 kV) which provides approximately 14% margin and apparently considers it close enough. On the 34.5 kV systems, many utilities are now putting in open tie protection and assuming that regardless of arrester characteristic or type of installation there is plenty of margin because there is no doubling.

Factors Affecting Margins

Neither the arrester discharge voltage nor the equipment BIL are constant. This section will review some of the items which reduce the margins we normally calculate.

Rate of Rise/Arrester Characteristics. The IR discharge of an arrester is found using the standard 8 x 20 μs wave. This wave has been used for many years and has been considered to be representative of lightning. Experimental data has indicated that actual lightning front times are much faster. Figure 5-22 showed data that is more representative of present industry thinking that front times of 2 microseconds or less are not unusual.

A time to crest of 1 microsecond would occur 17% of the time. If a wave with this rise time was impressed on an arrester, both the sparkover and the IR of the arrester would change from the values published for a standard 8 x 20 μs wave. Industry curves show that for such a wave the IR characteristic of a silicon carbide arrester increases approximately 30% and for an MOV the increase is about 10%. This is one of the main advantages of the MOV over the gapped silicon carbide arrester.

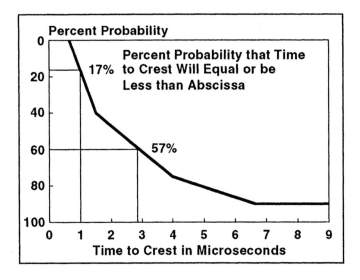

Figure 5-22. Times to Crest of Lightning

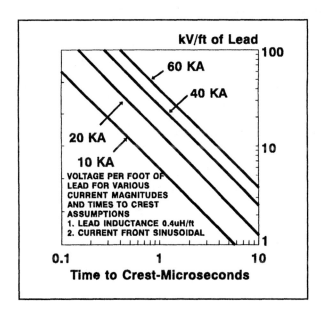

Figure 5-23. Voltage vs. Lead Length

201

Lead Length. The lead length of an arrester is the length of lead from the phase wire to the top of the arrester and the lead between the ground of the arrester and the metallic sheath of the cable. These leads produce a voltage due to the rate of rise of current passing through their inductance and added to the IR drop through the arrester in order to establish the total magnitude of discharge voltage impressed upon the system.

It has been established over the years that the drop is approximately 2 kV per foot based upon a lead inductance of 0.4 microhenries per foot, a 40 kA stroke, and an average rate of rise of 5 kA per microsecond.

Some time ago, tests were made using 7.5 feet of test lead and a current of only 9.5 kA crest and having a wave shape of 2.6 x 8 microseconds showing that the voltage appearing across the lead as 24.6 kV or 3.28 kV per foot. It is not always understood that the initial rate of rise of current is usually much faster than the average rate. Even assuming that the wave front is sinusoidal (and it is faster than sinusoidal) the initial rate of rise can readily be shown to be 1.57 (crest-current/time to crest). Figure 5-23 shows the arrester voltage per foot of lead length for various currents and times to crest assuming a sinusoidal current and a lead inductance of .4 microhenries/foot.

While the ANSI guides do point out that the arrester discharge voltage should be considered to be the total of the arrester IR and the lead voltage, it is rare when margins calculated for underground protection include this. Also, C62.2-1981, suggests that the acceptable voltage per foot of lead length is 1.6 kV/ft. which might be considered low.

BIL Deterioration. The deterioration of BIL in transformers and cables has received considerable attention in recent years primarily due to a much higher than expected failure rate. About 10 years ago, a research project sponsored by ERDA studied the effect of aging and loading on distributor transformer BILs. The units tested were from several manufacturers and rated 25 kVA, 95 kV BIL, with a 65°C-rise insulation system. The diagram shown below in Figure 5-24 indicates the results of these tests for the aged units.

Some of the general conclusions drawn from this study are:

- Almost 50% of the aged units failed below the 95 kV level.
- Aging and loading reduced initial BIL to an average of 64%.

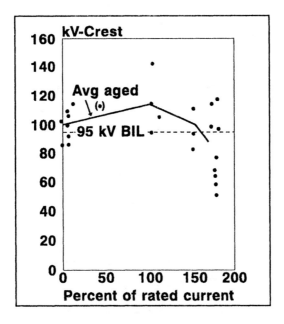

*Figure 5-24. BIL Deterioration vs.
Aging and Loading*

Reflections. Reflections at the open point can be reduced by using open tie arresters but are not eliminated. Studies indicate that the maximum voltage that can be seen on an underground cable using gapped silicon-carbide for riser pole and open tie protection is equal to IR discharge + 1/2 sparkover. Since sparkover (S.O.) is approximately equal to IR at 10 kA, the maximum voltage is approximately 150% of IR. The initial thought was that MOVs, having no sparkover, would eliminate this problem. Such is not the case. Studies performed with MOVs at the open point indicate that reflections with peaks of approximately 130% of IR are typical. If the effect of the increased rate of rise at the open tie is factored in, these estimates may be even higher.

Other. There are several other areas which also contribute to the uncertainty of the protective level. For example, while most utilities calculate margins based on a 10 kA discharge current and the more conservative on a 20 kA discharge, there is certainly a considerable body of evidence to suggest that the discharge currents could be considerably higher. Also the effect of system voltage polarity, effectiveness of grounding, multiple taps, etc., can all have a major effect on protective margins.

Considerations When Calculating Margins

The present use of the 20% margin factor does not properly address all the factors just discussed. Since lightning is a variable, so also are the degrees of protection afforded the utility equipment. It is necessary and possible, however, to address the concerns of varying arrester characteristics, lead length, BIL deterioration, reflections, magnitudes, etc., and ascertain whether these factors can all reasonably be accounted for within the 20% value. While guidelines mention many of the items discussed, they are unclear as to methods of quantifying their effects.

Table 5-8, shown below, is an attempt to put values on various degrees of concern. For example, a utility having a 13.2 kV system located in a low isokeraunic level area and with a low historical failure rate might fit into the "small concern" category. A utility using 34.5 kV in a high isokeraunic level area and experiencing a high failure rate problem might fit into the "extreme concern" category.

Table 5-8. Change in Arrester BIL Characteristics				
	Arrester Characteristics Change Rate of Rise	Lead Length kV	BIL Deterioration	Open-Tie Arrester Reflection Coefficient MOV
Small Concern	5%	3.2	-10%	1.3
Moderate Concern	10%	8	-20%	1.3
Extreme Concern	15%	24*	-30%	1.3
*6 ft. at 4 kV ft. = 24 kV				

As an example, suppose we are trying to protect a 34.5 kV, 4-wire, underground system using 27 kV distribution class MOV arresters at the riser pole and at the open tie. Figure 5-25 illustrates the system.

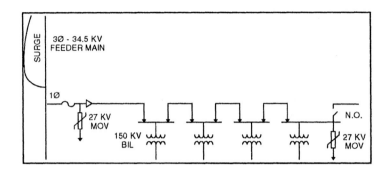

Figure 5-25. 34.5 kV Underground Lateral

Some standard parameters used in the calculation of margin are as follows:

BIL   - 150 kV
IR @ 20 kA   - 100 kV
Lead Voltage - 6.4 kV (4 ft. of leads at 1.6 kV ft.)

The most common method of calculating this margin is to assume that the open tie arrester prevents doubling and the lead length voltage is minimal. This calculation would be as follows:

$$\% \text{ Margin} = \frac{\text{BIL} - \text{IR}}{\text{IR}} \times 100\% = \frac{150 - 100}{100} \times 100\% = 50.0\% \quad \text{5-7}$$

This margin is, of course, almost double the value of the guidelines and might be interpreted as providing a very good level of protection.

ANSI Guidelines do suggest that lead length voltage be considered as part of the IR discharge. If this is done, the new margin can be calculated as follows:

$$\% \text{ Margin} = \frac{150 - (100 + 6.4)}{(100 + 6.4)} \times 100\% = 41\% \quad \text{5-8}$$

So, as we can see, commonly used methods produce margin which would seem to provide very adequate margins of protection.

205

If, however, we considered the items illustrated in Table 5-8, we can begin to see why our assumed margins may be too optimistic. If we factor in the three levels of concern, we can calculate the following margins:

- Small

$$\text{New BIL} = 150 \times .9 = 135$$

$$\text{New IR} = (100 \times 1.05 + 3.2) * 1.3 = 141 \text{ kV} \qquad 5\text{-}9$$

$$\% \text{ Margin} = \frac{135 - 141}{141} \times 100\% = -4\%$$

- Moderate

$$\text{New BIL} = 150 \times .8 = 120$$

$$\text{New IR} = (100 \times 1.10 + 8) * 1.3 = 153 \text{ kV} \qquad 5\text{-}10$$

$$\% \text{ Margin} = \frac{120 - 153}{153} \times 100\% = -22\%$$

- Extreme

$$\text{New BIL} = 150 \times .7 = 105$$

$$\text{New IR} = (109 \times 1.15 + 24) * 1.3 = 181 \qquad 5\text{-}11$$

$$\% \text{ Margin} = \frac{105 - 181}{181} \times 100\% = -42\%$$

Illustrated graphically in Figure 5-26, we can see the great disparity in margin calculation depending on what conditions are factored into the calculation.

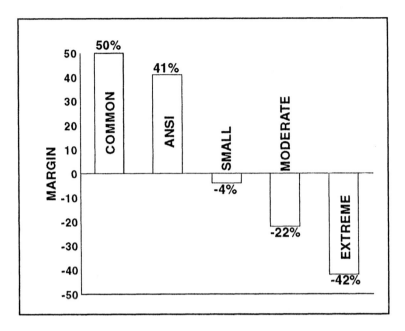

Figure 5-26. Margins of Protection at 34.5 kV

Effect of Traveling Waves

The previous section dealt with conditions which are widely discussed and generally accepted as considerations in a complete evaluation of insulation coordination. There are a number of other items, resulting from the fact that lightning is a complex traveling wave, that a good protection engineer should be aware of, which may explain the cause of some failures where protection was thought to be adequate.

Figure 5-27 shows a 12.47 kV system with a lateral underground tap of 400 meters. As was shown previously, the overvoltage protection of this system is normally considered adequate by most utilities with the use of one arrester at the riser pole. In many cases, the arrester at the riser pole is a "riser pole class" (non-official designation) arrester. These arresters have better characteristics than distribution class arresters and as such give even more margins. This, of course, tends to support the argument that only one arrester is necessary.

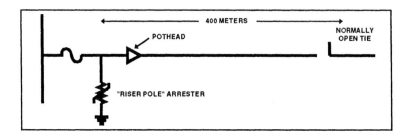

Figure 5-27. 12.47 kV Underground Lateral Tap

System conditions, lightning waveshapes, and the fact that lightning is a traveling wave, complicate any analysis. Add to this the fact that the arrester is not perfect either and a strong case can be made that few underground systems have adequate protection. The following are some examples of traveling wave phenomena which decrease protective margin. For all examples, a 1 x 20 μsec surge is used in the simulation since it is much more representative of actual lightning than the standard 8 x 20 μs. The arrester at the riser pole is a "riser pole class" with a 30 kV IR discharge for a 10 kA discharge.

TRAVELING WAVES

Voltage Doubling

Figure 5-28 shows the discharge voltage of the 9 kV riser pole arrester. This IR discharge is equal to approximately 30 kV and approximately a dc discharge. This waveshape propagates down the 400 meter long cable where it is reflected at the open tie point causing a doubling of the voltage to 60 kV as is seen in Figure 5-29. This doubling of discharge voltage would still allow a margin of approximately 58% if some of the considerations like BIL deterioration, lead length, etc., are neglected.

Addition of an open tie arrester to this system will prevent doubling at the open tie as can be seen in Figure 5-30. While this form of protection is very effective at the open point, it does not prevent all reflections. Figure 5-31 shows the cable midpoint voltage for this condition. As can be seen, positive waves are reflected back which add to the incoming voltage and produce a maximum voltage of about 40 kV or a margin of over 137%.

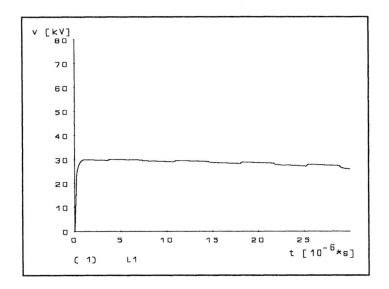

Figure 5-28. Riser Pole Voltage

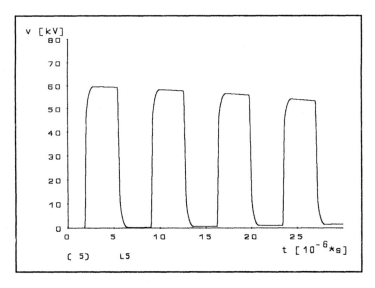

Figure 5-29. Cable Endpoint Voltage

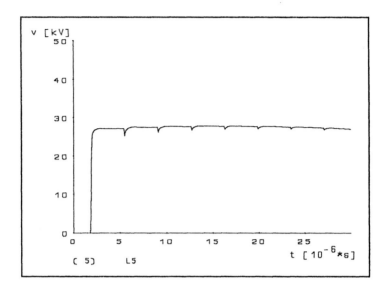

Figure 5-30. *Cable Endpoint Voltage With Open Tie Protection*

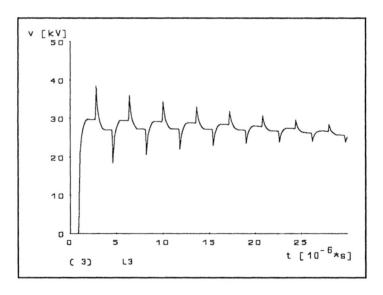

Figure 5-31. *Cable Midpoint Voltage*

Negative Trapped Charge

If a positive lightning stroke hits a distribution line at the precise time, the system voltage is at a negative peak $((12.47/\sqrt{3})*\sqrt{2} = 10.18\,\text{kV})$, the arrester will not go into complete conduction until the voltage impulse has compensated for the negative system voltage of -10 kV and the 30 kV normal discharge voltage. This equivalent impulse produces a 40 kV traveling wave at the riser pole as is shown in Figure 5-32.

This voltage will, of course, be reflected at the open point to about 70 kV and effectively reduce the margin to about 35%. While it could be argued that this margin is considerably greater than the recommended 20%, it should be pointed out that if even some of the previous considerations are factored in, the 20% margin will not be attained.

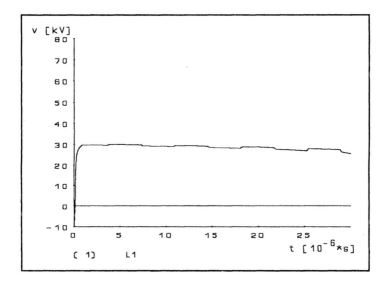

Figure 5-32. Riser Pole Voltage

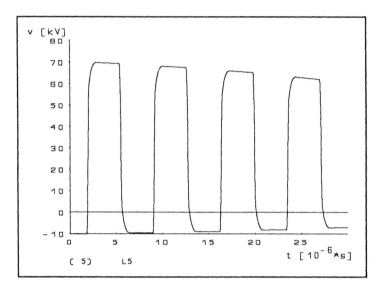

Figure 5-33. Cable Endpoint Voltage

Quadrupling

A number of lightning researchers have noticed that while the majority of lightning strokes are monopolar, some are bipolar, as shown in Figure 5-34. Phil Barker's paper "Voltage Quadrupling on a UD Cable" was the first to point out that if the timing of the bipolar wave and length of the cable were correct, voltage quadrupling could occur.

The wave of Figure 5-34 was simulated and discharged by the riser pole arrester. Figure 5-35 illustrates the discharge characteristic of the arrester showing the dramatic effect the nonlinearity of the arrester has on the shape of the traveling wave entering the cable. When this wave reaches the open point, it is reflected in a somewhat more complex manner resulting in a line-to-neutral voltage of approximately 120 kV (see Figure 5-36) which is over 25% greater than the system BIL.

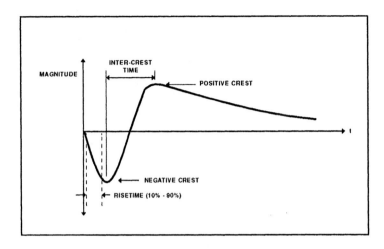

Figure 5-34. Bipolar Surge

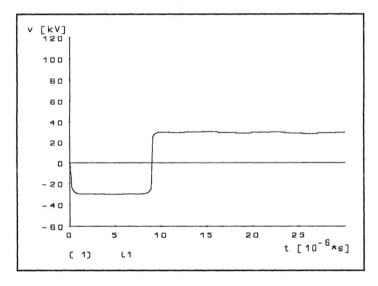

Figure 5-35. Riser Pole Voltage

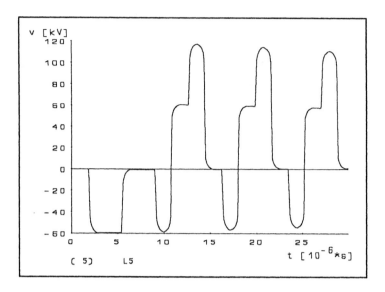

Figure 5-36. Cable Endpoint Voltage

Tapped Lateral

The tapped lateral is difficult to generally assess because the traveling wave pattern is so much more complex. For this example, two taps were added to a 200 meter cable, one 400 meters in length and the other 800 meters in length, as shown in Figure 5-37.

Under normal circumstances, the riser pole arrester voltage of 30 kV, would be doubled to 60 kV at the open point. Because of the more complex interaction of the two open points, voltage at the end of the longer 800 meter branch reached approximately 80 kV (see Figure 5-38) which gives less margin than the recommended 20%. Voltages at the tap point were almost 50 kV while voltages at the end of the shorter branch were as high as 60 kV (see Figures 5-39 and 5-40). While it is difficult to predict the maximum overvoltage seen on any tapped system, it is safe to say that the typical doubling philosophy should not be applied.

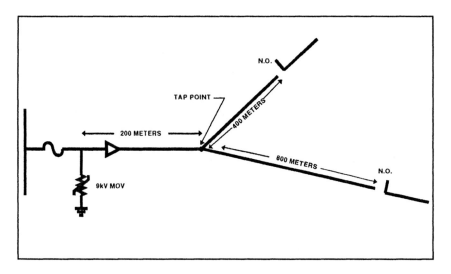

Figure 5-37.

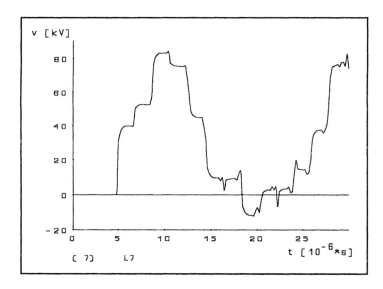

Figure 5-38. Voltage at the End of the Longer Branch

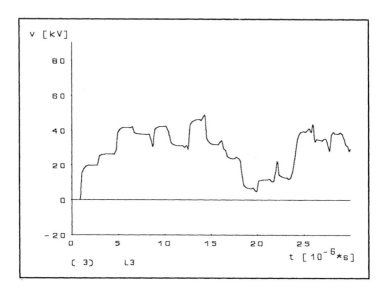

Figure 5-39. Tap Point Voltage

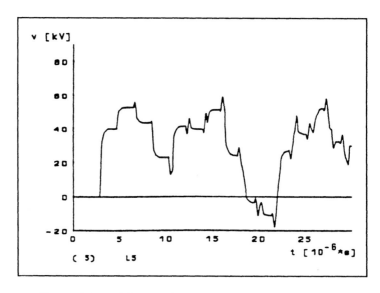

Figure 5-40. Voltage at the End of the Shorter Branch

Effect of Lateral Length

For simplicity, a traveling wave on an underground cable is typically shown as a "square wave" or in effect, a wave having an infinite "rate of rise". When this wave reaches the open point, it immediately doubles. The "riser pole" arrester has no effect on preventing this doubling, as was shown previously. While this technique simplifies the explanation of voltage doubling, it does not tell the entire story if the cable length is relatively short.

The front of the wave in reality looks more like a ramp or ramp function. As such, it takes time for it to reach its maximum value. For example the standard 8x20 μs wave, shown below, takes 8 microseconds to reach its crest. While this may seem like a very short time (and it is), from a traveling wave point of view it is quite long. If the assumption is made that the speed of the traveling wave is approximately 500 feet per microsecond then the wave can be thought of as cresting in about 4000 feet. For cables less than 2000 feet long, then the reflected wave has already returned before the crest current is discharged at the riser pole.

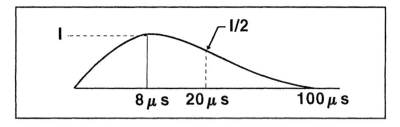

Figure 5-41

The surge reflection coefficient at the open point is positive because the open point represents an infinite impedance to the wave. On the other hand, the surge coefficient at the riser pole is negative since the riser pole arrester has a lower impedance than the surge impedance of the cable. This negative coefficient means that reflected waves will have their signs reversed (i.e., positive reflected waves become negative reflected waves and vice-versa). Figure 5-42, shown below, illustrates the interaction of traveling waves on short lines. As can be seen, the incident waves enters the cable. At this point, little if any arrester operation is taking place because the current level and consequent voltage level are still very low. The incident wave is reflected and adds to the incoming wave as shown in Figure 5-41. (It should be noted that the incoming wave is still low in magnitude). The reflected wave is now re-reflected due to the negative coefficient of the arrester and in effect lowers the voltage on the cable because it is negative as shown in Figure 5-42. The process continues as the incident waves increases in magnitude. The net effect, however, is that while the voltage does indeed increase above the discharge level of the arrester, it does not double.

While some engineers maintain that due to the above phenomena, short cables (the shorter the better) do not need open tie protection, this may not be completely correct. The 8 x 20 µs wave is not truly representative of lightning. Lightning has much higher rates of rise and can be expected to reach a crest in one microsecond or even less. While a cable of 2000 feet or less might see some benefit with an 8 x 20 µs wave and be classified as a "short line", the same cable would see the full doubling effect from a 1 x 20 µs wave and be classified as a "long line".

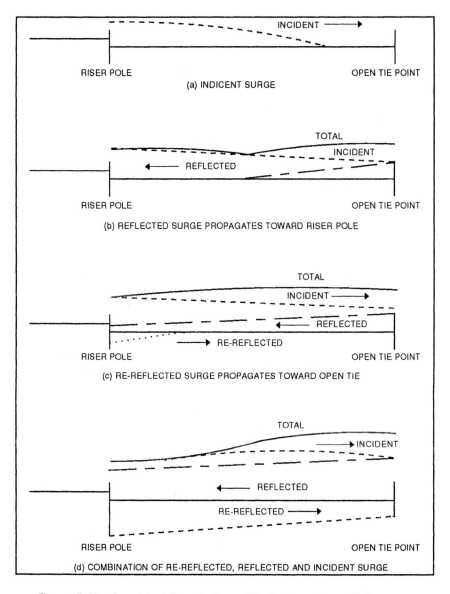

Figure 5-42. Graphical Description of Reflection Cancellation on a Short Line

Simulations have shown that a cable would have to be about 200 feet or less (for a 1 x 20 μs wave) in length before it could begin to see any benefit attributed to a "short line".

Summary of Recommendations

As has been demonstrated, considerations affecting underground protective margins are not constant. To summarize, some factors which greatly affect margins are as follows:

*   Effect of ground potential rise
*   Rate of rise on current
*   Lead length
*   BIL deterioration due to aging
*   Reflections (with and without open tie protection)
*   IR discharge changes due to aging
*   Bipolar waveshapes
*   Tapped laterals
*   Negative trapped charge.

While it is not possible to elaborate on these items for all situations, it is the author's opinion after years of analysis that the following general rules, shown in Table 5-9, for both straight and tapped laterals should be adhered to for proper underground protection.

LINE PROTECTION

Line Insulation

The primary insulation for distribution lines is air. Air, rather than other materials such as rubber or other solid dielectrics, is used because it is a self-healing dielectric. If a dielectric breakdown occurs (flashover) and the arc is de-energized, with a self-healing dielectric, the insulation is restored and the line can be re-energized without the insulation having to be repaired. With solid dielectrics, once failure occurs, the insulation must be repaired or replaced before the line can be re-energized.

Table 5-9. Recommended Arrester Locations		
Voltage	Feeder Configuration	Arrester Locations
15 kV	Radial	Riser Pole Open Tie
25 kV	Radial	Riser Pole Open Tie
35 kV	Radial	Riser Pole Open Tie Midpoint (near OT)
15 kV	Tapped Lateral	Riser Pole Tap Point
25 kV	Tapped Lateral	Riser Pole Tap Point
35 kV	Tapped Lateral	Riser Pole Tap Point All Open Ties

Lines must be supported in air and the supporting structures are usually electrically weaker insulators than the air. To maximize this insulation strength, porcelain or polymer insulators are used, often in conjunction with wood.

The voltage level at which flashover will occur on distribution structures is a function of the basic insulation level (BIL) of the structures. A direct lightning strike to a line will always strike the line not more than one-half span from a structure. For distribution lines, this means that the strike point will generally not be more than 200 feet from a structure. A flashover will generally occur at the structure closest to the strike. Flashover of multiple structures on both sides of the strike point are also common.

While accurate BIL for structures is obtained by testing the structure with a surge generator, estimates of BIL can be made. For distribution lines, wet flashover values for negative impulses are used and added directly to the impulse flashover value for wood. BIL for wood varies by type of wood but generally can be assumed to be about 100 kV per foot dry. Wet value is approximately 75 kV per foot. For a structure with a

100 kV BIL insulator and a 3 foot spacing of wood, the BIL would be approximately 325 kV.

The primary concern when designing structures is to attain a 300 kV or greater BIL level to ensure that only direct strikes to the line will cause a flashover. This is normally accomplished by utilizing the wood of the structure, itself. A standard 7.6/13.2 kV structure, which is shown in Figure 5-43, has a BIL of approximately 300 kV. This structure uses an 8 foot crossarm with wood crossarm braces; these wood braces are important in attaining the 300 kV BIL (note that if steel braces are used, part of the wood insulation is short circuited, resulting in a much lower BIL). The goal in designing distribution structures should be to equal or exceed the BIL of this structure.

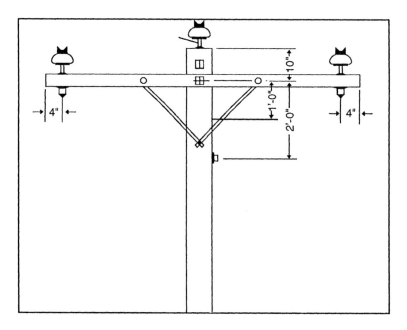

Figure 5-43. 3Ø Pole Head

Figure 5-44 shows a typical armless construction found in many newer areas. The design uses 20 kV insulators and has a BIL of approximately 150 kV.

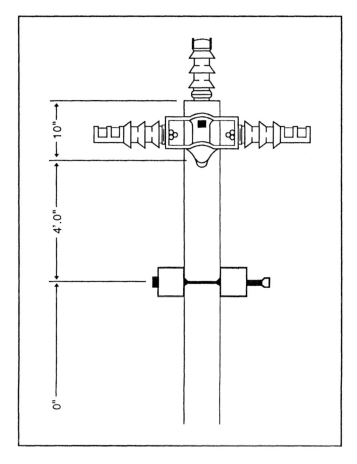

Figure 5-44. 3Ø Pole Head with 3Ø Tap Armless Construction 20 kV Insulators.

Types of Lightning Protection

No Protection. This type of protection is identified as one in which the line has exposed phase wires with the neutral, if used, located beneath the phase conductors. No shield wires or arresters are installed for line protection. As such, this type of line represents the worst case. Subtransmission lines with long runs may approach this type of construction and lack of lightning protection.

223

Distribution lines of this type will have equipments such as transformers, capacitor banks, reclosers, and regulators located along the line. This equipment will usually be protected by lightning arresters. These arresters will protect the equipment adequately, but will have only a modest effect on the flashover rate of the line itself. A distribution line having these equipments with associated lightning arresters would have a lightning outage rate approaching that for one of the types of lightning protection to be described below. The degree of protection would be proportional to the amount of equipment and the number of arresters.

Shield Wire. This type of protection is widespread in the transmission line area. On distribution lines it is used occasionally. It can be the system neutral conductor if this is located overhead and if it is grounded frequently along the line. When properly installed, the overhead ground wire will intercept nearly all lightning strokes which will terminate on the line. It is only effective, however, when ground footing impedances can be maintained at a low value.

Occasionally, a stroke will terminate on a phase conductor because of a shielding failure. The shielding failures are a function of the height above ground and of the shielding angle. For the conductor heights expected in the distribution area, shielding failures are not expected to be a problem provided the shielding angle is less than 40 to 50 degrees. The shielding angle is the angle (measured from the vertical) from the ground wire to the phase conductor.

Basically, as the impulse voltage between the two conductors (ground wire and nearest phase conductor) enters the breakdown range, a sheet of prebreakdown currents erupts between the two conductors. The total current reaches thousands of amperes if the conductors are of sufficient length, and causes a pronounced delay in breakdown by reducing the voltage somewhat in the same manner as a lightning arrester limits voltage by drawing a heavy current. Breakdown can still take place, but because of the voltage reduction by these prebreakdown corona currents, this breakdown may occur at 5 µs or more rather than at 1 or 2 µs which might be the classical value. In the meantime, because of this long delay in breakdown, reflections have the opportunity to arrive from nearby grounded poles and to reduce the voltage between conductors. A sufficient reduction in voltage within the breakdown time will prevent a flashover at the stricken location.

Arrester on Top Phase. This type of protection is not considered to be effective although it is quite common. The combination of top phase conductors and arresters act somewhat like an overhead ground wire system in which the footing resistance is increased by the amount of the arrester resistance. Shown below is a simplified one-line of a pole having two isolated conductors. One of the conductors (top phase) is protected by a lightning arrester.

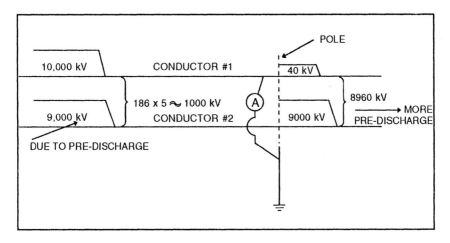

Figure 5-45. 13.2 kV System

Now assume that a lightning stroke hits the line. This stroke will cause a voltage of the same polarity in adjacent lines due to pre-discharge currents. The pre-discharge voltage is 186 kV per foot; so for a 5 foot spacing we would have an approximate 1,000 kV voltage difference between these two lines. For this example, it can be assumed that the top phase surge voltage is 10,000 kV and the conductor 2 voltage is 9000 kV or 10,000 kV - 1000 kV. Since the effective voltage difference between the lines has been reduced now to 1000 kV due to pre-discharge, there is less chance of flashover out on the line.

When the two surges get to the pole where the top phase is protected, the arrester will spark over and reduce the top phase voltage to approximately 40 kV for a 13.2 kV system. The lower phase however is still 9000 kV so the potential difference is now 9000 - 40 or 8960 kV making the probability of sparkover greater. Of course this situation will change with time because pre-discharge will take place again until the

225

next arrester location where the process repeats itself at a lower energy level.

The next figure shows the same situation with arresters on the two conductors. Here it can be easily seen that when the surges arrive at the pole both arresters will sparkover and the transmitted voltages to the next pole will be very low. This situation greatly reduces the probability of flashover.

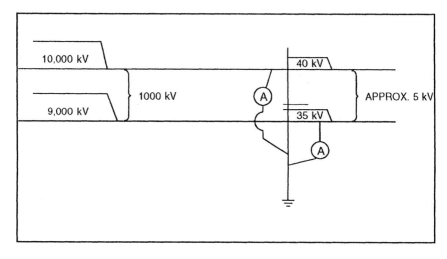

Figure 5-46

Arresters on All Phases. Prebreakdown currents have the effect of reducing the probability of flashover out on the span and of increasing the voltage stress at the poles having grounds for the overhead ground wire. In addition, it is noted that the voltage at these poles appears across the insulation of the conductors not involved in the stroke. It would seem likely, therefore, that arresters on all phases would eliminate lightning flashovers at the poles having arresters, and result in a significant reduction in the total number of flashovers.

With arresters installed on all phases, their ground connections would be tied together near the top of the pole and a common ground connection installed to ground at the foot of the pole. The traveling wave voltages would be similar to those seen for the previous figure illustrating two conductors and two arresters. The arresters, on all phases, must be located on the same pole to insure that the voltage waves rise and fall together, thus minimizing voltage stress between conductors. It is interesting to note that this scheme is immune to the effect of poor footing resistances since the entire line will rise and fall together and the arrester voltage which is between each conductor and the top of the pole, (approximately 40 kV) will be relatively constant. This is unlike the shield wire where a high footing resistance can cause the shield wire to have a higher voltage than the phase wire and a "backflash" can occur.

Comparison of Line Protection Schemes

Today it is possible to evaluate the performance of specific line designs with the use of digital simulations. These simulations can evaluate the effect of shield wires, arresters on each phase, spacing, rate-of-rise of current, grounding, neutral placement and other important parameters. Figure 5-47, shown below, is a comparison of some of these parameters for a 13.2 kV system. As can be seen, the best protective scheme is usually the use of arresters on all phases and the worst is typically having an arrester on the top phase only. The effect of line BIL has also been evaluated and illustrates that for very effective protection, a BIL of 300 kV or above is prudent regardless of the scheme used.

The types of construction used will be called flat, triangular, and post. Each type of construction has a probability of flashover which is a function of the line BIL. Figures 5-48, 5-49, and 5-50 illustrate the probability of flashover as a function of line BIL and arrester (ground) spacing.

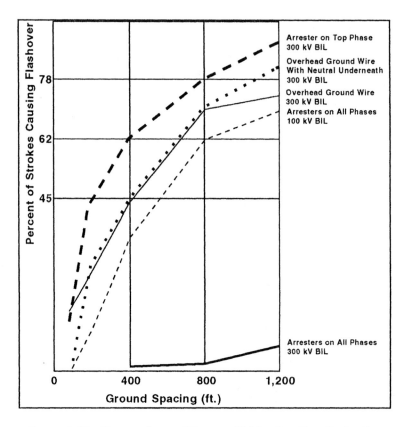

Figure 5-47. Comparison of Various Distribution Line Protective Measures

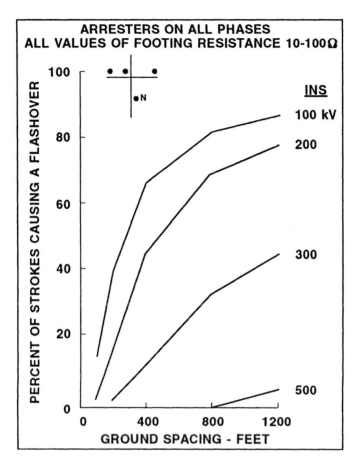

Figure 5-48. Flat Construction

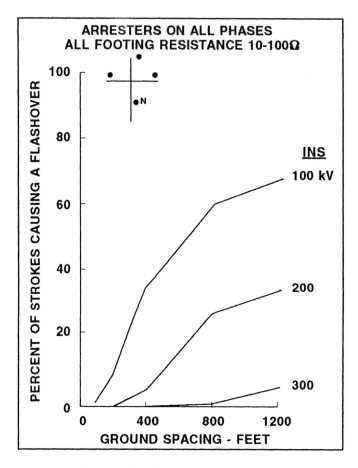

Figure 5-49. Triangular Construction

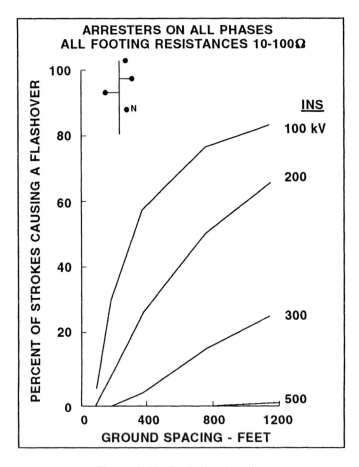

Figure 5-50. Post Construction

INDUCED STRIKES

There is a considerable amount of evidence to suggest that many of the flashovers occuring on an overhead distribution line might be the result of induced strikes, (i.e., nearby strikes that do not hit the line directly). The voltages caused by these strokes are considerably lower than those for direct hits (usually less than 300 kV). The rise times are also less, making spacing of arresters less critical.

Protection for flashovers from induced strikes can, in some cases, become more effective by simply increasing the BIL of the line design. This is normally done by increasing the insulator ratings, changing the position of the neutral wire, increasing the phase wire spacing, etc. Arrester protection having spacing every quarter mile or so, while not very effective on a direct hit, can be effective for induced strikes.

REFERENCE

1. P.P. Barker, "Voltage Quadrupling on a UD Cable", *IEEE Transactions on Power Delivery*, January 1990, Vol. 5 No. 1, Transmission & Distribution Conference, April 1989.

QUESTIONS

1. What is the average duration of a lightning stroke?

2. Plot a 10 kA lightning stroke on a 10 kA fault. What does this tell you about energy content of lightning?

3. Why is the 8 x 20 μs wave not particularly representative of lightning?

4. Name 4 conditions that increase the possibility of ferroresonance.

5. What is restrike?

6. Why does "current chopping" create overvoltages?

7. If a silicon carbide arrester did not have a gap, how much current (approx.) would it draw at nominal voltage?

8. Why don't MOVs require gaps?

9. What is the major difference between the 3 classes of arresters?

10. Define MCOV and TOV.

11. Why are arrester ratings for delta systems higher than for 4-wire multigrounded systems of same voltage?

12. Why is the TOV curve virtually useless to distribution engineers?

13. Select arrester rating for 34.5 kV, 4-wire multigrounded system.

14. What is the relationship between arrester rating and MCOV?

15. Arrester rating is based on what?

16. What is "insulation coordination"?

17. What's one disadvantage of externally gapped arresters?

18. Do MOVs give more "margin" than gapped silicon carbide? Explain.

19. Calculate margin for an overhead 34.5 kV, 4-wire multigrounded system.

20. Why do margins go down as system voltage level goes up?

21. Why are margins for underground less?

22. Why don't we usually worry about doubling of voltage (due to traveling waves) on overhead systems?

23. What is the margin for a 34.5 kV URD system having only a rise pole arrester and a BIL of 125 kV?

24. Why can the arrester lead length be so detrimental to protective margins?

25. Why are protective margins much lower than we think?

26. Name 3 ways to improve URD performance?

27. Explain effect of "trapped charge" on URD protection?

28. What type of wave must you have to obtain quadrupling on a URD system?

29. It is well known that use of a single riser pole arrester can be very effective for "short lines". What determines when a line is considered short?

30. How many arresters should be used for 34.5 kV URD? 13.8 kV?

6

DISTRIBUTION RELIABILITY

INTRODUCTION

The great majority of service interruptions the customer sees are caused by problems on the distribution system. The overhead system is primarily affected by meteorological conditions such as wind and lightning whereas the underground system, while it sees fewer outages, is particularly concerned with service restoration time following a failure.

It is interesting to note that in recent years, with the emphasis on power quality to more sensitive loads, the meaning of system reliability is changing. For example, the momentary interruption caused by the reclosing breaker to clear a temporary fault was only a few years ago considered to be a minor annoyance. Today, this same operation can cause havoc in the form of computer shutdowns, industrial process shutdowns and a multitude of blinking clocks.

FUNDAMENTALS

Calculating Reliability

Reliability is defined as "the probability of a device performing its purpose adequately for the period of time intended under the operating conditions encountered". This definition implies that reliability is the probability that the device will not fail to perform as required for a certain length of time. The problem with this definition when it is applied to utility service is that it does not consider the duration of the failure. It is conceivable that a "bare bones" distribution design (i.e., one with no switches, fuses, etc.) could have a very high reliability (by this definition, anyway) but would, of course, have unacceptable repair time.

Reliability of a distribution system is really evaluated in terms of outage rate and outage duration. The basic formula for evaluating a radial utility distribution system is as follows:

Components in Series

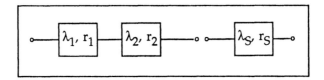

Figure 6-1. Components in Series

A system consisting of two components in series with outage rates λ_1 and λ_2 and repair times r_1 and r_2 respectively has the following reliability indices.

System outage rate,

$$\lambda_S = \lambda_1 + \lambda_2 \qquad\qquad 6\text{-}1$$

System average outage duration,

$$r_S = \frac{\lambda_1 r_1 + \lambda_2 r_2}{\lambda_1 + \lambda_2} = \frac{\Sigma\lambda r}{\Sigma\lambda} \qquad\qquad 6\text{-}2$$

System average total outage time,

$$U_S = \lambda_S r_S \qquad\qquad 6\text{-}3$$

Components in Parallel

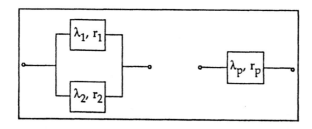

Figure 6-2. Components in Parallel

235

A system consisting of two components in parallel has the following reliability indices.

System outage rate,

$$\lambda_p = \lambda_1 \lambda_2 (r_1 + r_2) \qquad\qquad 6\text{-}4$$

System average outage duration,

$$r_p = \frac{r_1 r_2}{r_1 + r_2} \qquad\qquad 6\text{-}5$$

System average total outage time,

$$U_p = \lambda_p r_p \qquad\qquad 6\text{-}6$$

Example No. 1

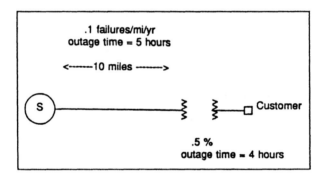

Figure 6-3.   Components in Series

Calculate:

1. Customer outages per year
2. Customer average outage duration
3. Customer average total outage time
4. Conclusions

Solution No. 1

$$\lambda_1 = 1.0 \qquad r_1 = 5$$
$$\lambda_2 = .005 \qquad r_2 = 4 \qquad\qquad 6\text{-}7$$

$$\lambda_S = 1.0 + .005 = 1.005$$

$$r_S = \frac{(1)\,(5) + (.005)\,(4)}{1 + .005} = \frac{5 + .02}{1.005} = 4.995 \qquad 6\text{-}8$$

$$U_S = \lambda_S\, r_S = 1.005\,(4.995) = 5.02$$

Example No. 2

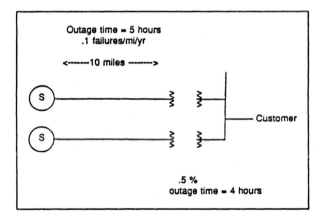

Figure 6-4. Components In Parallel

237

Solution No. 2

$$\lambda_1 = \lambda_2 = 1.005$$

$$r_1 = r_2 = 4.995$$

6-9

$$\lambda_P = \lambda_1 \lambda_2 (r_1 + r_2)$$

$$= (1.005)(1.005) \frac{(4.995 + 4.995)*}{8760}$$

6-10

$$= \frac{(1.01)(9.99)}{8760} = \frac{1.01}{8760} = .001 \text{ outages/year}$$

$$r_P = \frac{(4.995)(4.995)}{4.995 + 4.995} = 2.5$$

6-11

Reliability Indices

There are two methods of calculation of reliability indices in the U.S. The majority of companies calculate indices based on the number of customers per outage and the duration of the outage. There is another group that calculates indices based on the amount of load that is lost. They do not keep track of the actual load that is lost, but rather use the peak load value of any given section in calculating their indices. The companies that base their reliability indices on lost load are doing so as a precursor to the customer based index system. At the present time, their data bases are not as extensive as the other companies.

Indices have been defined by various groups such as the IEEE (Institute of Electrical and Electronics Engineers), EEI (Edison Electric Institute), EPRI (Electric Power Research Institute) and CEA (Canadian Electric Association) as follows:

SAIFI - System Average Interruption Frequency Index

$$\text{SAIFI} = \frac{\text{Total No. Customers Interrupted}}{\text{Total No. Customers}}$$
$$= \frac{(\text{No. Customers Interrupted}) * (\text{No. of Interruptions})}{\text{Total No. Customers}} \quad \text{6-12}$$

SAIDI - System Average Interruption Duration Index

$$\text{SAIDI} = \frac{\sum \text{Customer Interruption Durations}}{\text{Total No. of Customers}}$$
$$= \frac{\sum (\text{Duration of Outage}) * (\text{No. Customers Affected})}{(\text{Total No. Customers})} \quad \text{6-13}$$

CAIFI - Customer Average Interruption Frequency Index

$$\text{CAIFI} = \frac{\text{Total No. Customer Interruptions}}{\text{No. of Customers Affected}} \quad \text{6-14}$$

CAIDI - Customer Average Interruption Duration Index

$$\text{CAIDI} = \frac{\sum \text{Customer Interruption Durations}}{\text{Total Number of Customer Interruptions}} \quad \text{6-15}$$

ASAI - Average Service Availability Index

$$\text{ASAI} = \frac{\text{Customer Hours Service Availability}}{\text{Customer Hours Service Demand}} \quad \text{6-16}$$

ATPII - Average Time until Power Restored

$$\text{ATPII} = \frac{\sum \text{Interruption Duration}}{\text{No. of Interruptions}} \quad \text{6-17}$$

CMPII - Customer Minutes per Interruption

$$CMPII = \frac{\sum \text{Customer Minutes per Duration}}{\text{No. of Interruptions}} \qquad 6\text{-}18$$

The load based indices were developed by the PEA (Pennsylvania Electric Association) and are defined as shown below in Equations 6-19 through 6-22.

These indices are based on the peak connected load during an interruption rather than on the number of customers interrupted. These indices are used due to the limitations of the companies data base. The companies that are using this system plan to upgrade their systems to customer based indices when their customer information systems track the number of customers interrupted. Typical numbers for customer based indices are shown in Table 6-1.

ASIDI - Average System Interruption Duration Index

$$ASIDI = \frac{\text{KVA Minutes Interrupted}}{\text{Total Connected KVA Served}} \text{ (Minutes)} \qquad 6\text{-}19$$

ASIFI - Average System Interruption Frequency Index

$$ASIFI = \frac{\text{KVA Interrupted}}{\text{Total Connected KVA Served}} \text{ (Avg. Intrpt.)} \qquad 6\text{-}20$$

ACIDI - Average Circuit Interruption Duration Index

$$ACIDI = \frac{\text{KVA Minutes Interrupted}}{\text{Total Connected KVA Served}} \text{ (Minutes)} \qquad 6\text{-}21$$

ACIFI - Average Circuit Interruption Frequency Index

$$ACIFI = \frac{\text{KVA Interrupted}}{\text{Total Connected KVA Served}} \text{ (Avg. Intrpt.)} \qquad 6\text{-}22$$

Table 6-1. Customer Based Indices			
SAIDI	**SAIFI**	**CAIDI**	**ASAI**
95.9 min/yr	1.18 int/yr	76.93 min/yr	.999375 int/yr

Survey Results

In 1990, a survey was conducted of utility distribution reliability practices throughout the United States. Among the most pertinent findings were that the most commonly used indices are ASAI, SAIFI, SAIDI, and CAIDI as shown in Figure 6-5. In less technical terms, most utilities are interested in the following:

Average Customer Minutes Outage per Year. This is the average cumulative amount of time a given customer would expect to be without service in a one year period (CMO or SAIDI). A typical value for SAIDI is about 100 minutes.

Average Duration of a Given Outage. Some outages can be very short, others quite long. This number is simply the average duration (CAIDI). A typical number for CAIDI is about 80 minutes.

Average Annual Number of Outages. This is the average number of times a given customer can expect an outage in a year (SAIFI). The average customer in the U.S. can expect to see about 1.20 outages per year (SAIFI = 1.20).

Availability of Service. This is the ratio of the total number of customer hours that service was available to the total customer hours demanded (ASAI). Although rarely used in discussion of distribution reliability, it is generally calculated for comparison to other parts of the utility system.

Another important survey finding was that most utilities do not classify an interruption as an "outage" until its duration exceeds about 5 minutes (see Figure 6-6). This of course means that temporary faults and their associated breaker or reclosing operations are not considered outages.

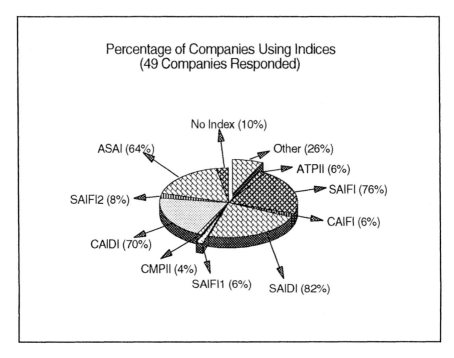

Figure 6-5. Most Commonly Used Indices

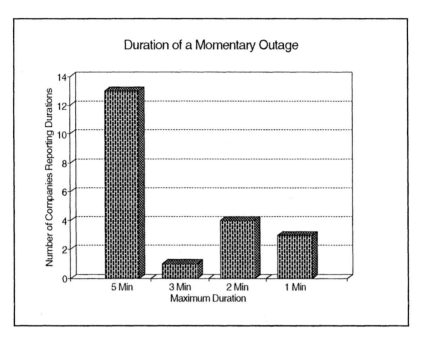

Figure 6-6. Typical Durations of Momentary Outages

Causes of Outages

It is very difficult to generalize on the causes of outages since they are somewhat particular to the geography of the region and the practice of the utility. It is easy to assume, for example, that utilities in high lightning areas will see a large number of outages caused by lightning. This, however, may not be true since many of these same utilities use overvoltage protection practices which greatly mitigate the effects of lightning. Figure 6-7, shown below, illustrates results from a recent study. As can be seen, a large percentage of outages cannot be identified. The most common reported causes of outages are lightning and tree contact. One utility in this same study reported that virtually all their outages were animal related.

243

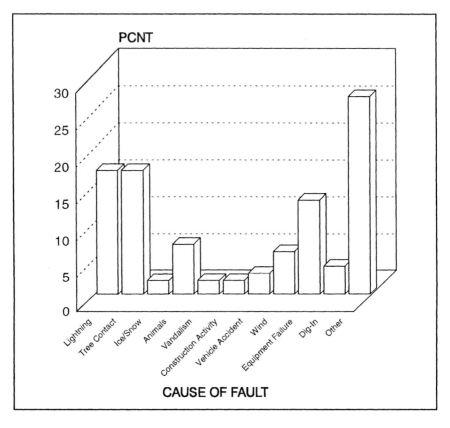

Figure 6-7.

Failure Rates

The largest problem in performing a reliability study is determining failure rates. Because of the newness of many system designs being studied and their related equipment, there is many times very little data available. Good failure rate data takes many years to develop. For example, it may take 10 or more years before a problem such as corrosion affects system operation. In coming up with failure rates to be used in a study, a subjective judgement based on the data available, along with considerable advice from utility engineers, in the specific area to be studied should be used. The failure rates shown in the table are thought to be quite representative of what a typical utility would experience.

244

Table 6-2	
Circuit Element	**Failure Rate**
Primary Underground Cable	0.03
Secondary Underground Cable	0.11
Service Underground Cable	0.11
Cable Terminations	0.002
Distribution Transformer	0.003
Transformer Fuse	0.005

The second problem with regard to failure rate data is to determine how a failure rate is allocated to its failure modes. For example, a primary cable can have a failure caused by the cable opening or a cable shorting to ground. Generally, all modes of failure are included when a failure rate is given. Sometimes, only the predominant mode is included. For primary cable and terminations, the failure mode for the vast majority of cases is a short to ground. On the secondary cable, past information shows that if the cable shorts, it usually burns itself clear (thus creating an open); thus, the dominant failure mode for secondary cable is the "open" condition. Table 6-3 shows the failure rates with both the predominant mode of failure and the secondary mode.

Table 6-3. Standard Failure Rates			
Circuit Element	**Total Failure Rate/yr.**	**Failure Mode**	**Failure Rate/yr.**
A. Secondary U.G. Cable	.11	Open Short	.099 .011
B. Dist. Transformer	.003	Open Short	0 .003
C. Loop Switch	.005	Won't Close Short	0 .005
D. Service U.G. Cable	.11	Open Short	.099 .011
E. Cable Terminations	.002	Open Short	0 .002
F. Primary U.G. Cable	.03	Open Short	0 .03
G. Transformer Fuse	.005	Open Won't Blow	.005 .0

POWER QUALITY

Fault Selective Feeder Relaying (FSR)

The feeder relay is the primary device in the protection of the feeder main and also the protection of the lateral fuse during temporary fault conditions. If a feeder is underground, it is usually assumed that all faults are permanent in nature. On the more predominant overhead feeder, such is not the case and what's called a reclosing relay is used on the feeder breaker.

The reclosing relay recloses its associated feeder breaker at preset intervals after the breaker has been tripped by overcurrent relays. Survey results indicate that approximately 70 percent of the faults on overhead lines are non-persistent. Little or no physical damage results if these faults are promptly cleared by the operation of relays and circuit breakers. Reclosing the feeder breaker restores the feeder to service with a minimum of outage time.

If any reclosure of the breaker is successful, the reclosing relay resets to its normal position. However, if the fault is persistent, the reclosing relay recloses the breaker a preset number of times and then goes to the lockout position.

The reclosing relay can provide an immediate initial reclosure plus three time-delay reclosures. The immediate initial reclosure and/or one or more of the time-delay reclosures can be made inoperative as required. The intervals between time-delay reclosures are independently adjustable.

The primary advantage of immediate initial reclosing is that service is restored so quickly for the majority of interruptions that the customer does not realize that service has been interrupted. The primary objection is that certain industrial customers cannot live with immediate initial reclosing.

The majority of utilities use a three-shot reclosing cycle with either three time-delay reclosures or an immediate initial reclosure followed by two time-delay reclosures. In general, the interval between reclosures is 15 seconds or longer, with the intervals progressively increasing (e.g., a 15-30-45-second cycle), giving an overall time of 90 seconds.

Fault selective feeder relaying allows the feeder breaker to clear non-persistent faults on the entire feeder, even beyond sectionalizing or branch fuses, without blowing the fuses. (See Figure 6-8). In the event of a persistent fault beyond a fuse, the fuse will blow to isolate the faulty section. Operating engineers report reductions of 65 to 85 percent in fuse blowing on non-persistent faults through the use of this method of relaying.

247

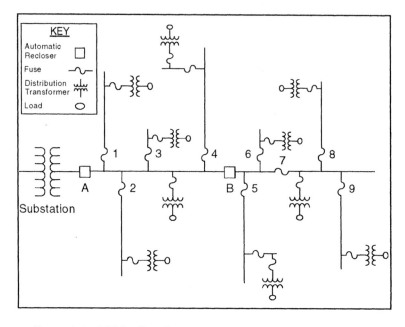

Figure 6-8. Distribution Feeder with Automatic Reclosers and Fuse Cutouts

Blocking the Instantaneous

The biggest drawback to "feeder selective relaying" is that the entire feeder sees an interruption every time a lateral experiences a temporary fault. With today's sensitive loads, especially in high-tech commercial areas, the high number of interruptions caused by successful reclosing operations to clear temporary faults can present more of a problem than fewer interruptions of much longer duration. It is for this reason that some utilities have decided to "block the instantaneous trip" and NOT save the fuse for temporary faults on the lateral. These, of course, now become permanent outages.

Many public service commissions (PSC) across the U.S. require utilities to report reliability indices. At the present time, most PSCs require information about permanent but not momentary outages. While changing the protection from FSR to not saving the fuse has the advantages just stated, the utility should be cognizant of the fact that by blocking the instantaneous trip, they will change their indices. Allowing

248

the fuse to blow on temporary faults will result in permanent outages for that section of line. Thus, this change will ALWAYS make their reported indices look worse to the public utilities commission. Even the utilities that calculate indices but do not report them calculate the indices based on permanent outages.

Figure 6-9 illustrates this point. As can be seen, while the (unreported) momentaries go down less than 25%, the reported outages go up over 50%.

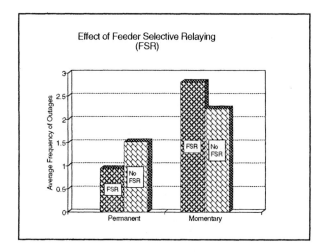

Figure 6-9.

Example No. 3

Assume you are the only customer on a substation that has the failure rate and outage duration characteristics shown in Figure 6-10. The PSC asks you to assess the effect adding a recloser to the mid-point of the feeder (See Figure 6-11). Calculate SAIFI, SAIDI, and CAIDI before and after the system change.

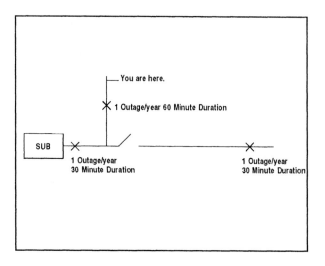

Figure 6-10. System Without Recloser

Solution No. 3

Before

$$\text{SAIFI} = \frac{\text{Total No. Customers Interrupted}}{\text{Total No. Customers}}$$

$$= \frac{(\text{No. Customers Interrupted}) * (\text{No. of Interruptions})}{\text{Total No. Customers}} \quad 6\text{-}23$$

$$= \frac{(1) * (3)}{1} = 3 \text{ per year}$$

$$\text{SAIDI} = \frac{\sum \text{Customer Interruption Durations}}{\text{Total No. of Customers}} =$$

$$\frac{\sum (\text{Duration of Outage}) * (\text{No. Customers Affected})}{(\text{Total No. Customers})} \quad 6\text{-}24$$

$$= \frac{30 * 1 + 30 * 1 + 60 * 1}{1} = 120 \text{ minutes}$$

$$\text{CAIDI} = \frac{\sum \text{Customer Interruption Durations}}{\text{Total Number of Customer Interruptions}}$$

6-25

$$= \frac{(30 + 30 + 60)}{3} = \frac{120}{3} = 40$$

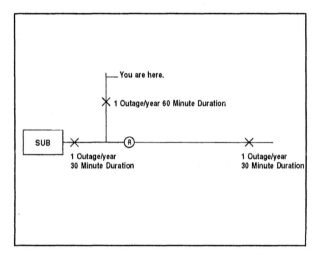

You are here.

✕ 1 Outage/year 60 Minute Duration

SUB ✕ ──(R)── ✕

1 Outage/year
30 Minute Duration

1 Outage/year
30 Minute Duration

Figure 6-11. System With Recloser

After

$$\text{SAIFI} = \frac{(1)\ (2)}{1} = 2/\text{year}$$

$$\text{SAIDI} = \frac{(60 \cdot 1) + (30 \cdot 1)}{1} = 90 \text{ minutes}$$

6-26

$$\text{CAIDI} = \frac{(30 + 60)}{2} = 45 \text{ minutes}$$

FACTORS AFFECTING RELIABILITY

Higher Voltages

The decision of a utility to standardize on a higher distribution voltage level is a very difficult one. While much of the decision is based on load density, availability of right-of-way, losses and substation placement, a major item which must be considered is reliability.

Higher voltage levels, because they increase exposure, decrease reliability. It can be argued that by increasing sectionalizing capability, higher voltage levels can be made as reliable as the lower voltages. This is easier said than done as is illustrated below:

Suppose a utility operating at 13.2 kV decides to convert to 34.5 kV and extend its feeder by 260% (34.5/13.2 x 100%) as shown below. Assuming the 13.2 kV system served 1000 customers, it can be assumed that with the same load density (and lateral length) the 34.5 kV system can serve 2600 customers.

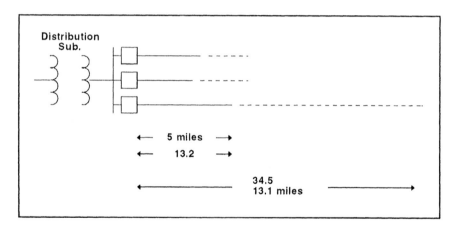

Figure 6-12

Assume:

CMO (on a feeder) = feeder failure rate · repair time · # of customers
$$= f \cdot m \cdot r \cdot n$$

where f = feeder failure rate/mile
      m = miles
      r = repair time (minutes)
      n = number of customers

and that the feeder failure rate is .1/miles/year and the repair time is 120 minutes. The CMO's for the two designs can be calculated as:

@ 13.2

$$CMO = .1 \cdot 5 \cdot 120 \cdot 1000 = 60,000 \text{ minutes/year}$$
6-27

$$\Rightarrow \text{average customer minutes outage} = \frac{60,000}{1,000} = 60 \text{ minutes}$$

@ 34.5

$$CMO = .1 \cdot 13.1 \cdot 120 \cdot 2600 = 408,720 \text{ minutes/year}$$
6-28

$$\Rightarrow CMO \text{ (average)} = 157.2 \text{ minutes}$$

If we attempt to make 34.5 kV more reliable by adding a midpoint switch, the CMO is calculated as follows:

$$CMO = .1 \cdot 6.6 \cdot 120 \cdot 2600 + .1 \cdot 6.6 \cdot 120 \cdot 1300$$

ie., CMO (feeder) = 205,920 + 102,960 = 308,880 customer minutes

$$\text{Average CMO} = \frac{308,880}{2600} = 118.8 \text{ minutes}$$
6-29

Dividing the feeder into 3 parts; we calculate feeder CMO as follows:

$$CMO = .1 \cdot \frac{13.1}{3} \cdot 120 \cdot 2600 + .1 \cdot \frac{13.1}{3} \cdot 120 \cdot 1733.3$$

$$+ .1 \cdot \frac{13.1}{3} \cdot 120 \cdot 866.7$$

6-30

$$= 136,240 + 90,824.9 + 45,415.1$$

$$= 272,480 \text{ minutes/year}$$

$$\text{Average CMO} = \frac{272,480}{2600} = 104.8 \text{ minutes}$$

∴ since even 3 sections of 34.5 kV cannot approach the 60 minute criteria of the 13.2 kV, it could be concluded that similar reliability can be difficult to achieve.

System Design

A three-phase feeder main can be fairly short, on the order of a mile or two, or they can be as long as 30 miles. Voltage levels can be as high as 34.5 kV, with the most common voltages being in the 15 kV class. While most of the 3-phase mains are overhead, much of the new construction, particularly the single-phase lateral construction, is being put underground. Underground systems have the advantage of immunity from certain types of temporary fault conditions like wind, direct lightning strikes, animals, etc. Permanent faults, on the other hand, are much more difficult to locate and repair and have been the subject of much concern in recent years. The more complex and costly distribution designs shown in Chapter 1 have certain advantages in regard to reliability. Each of the five systems previously described can be evaluated in terms of reliability for traditional loads as shown below. As can be seen, in Figure 6-13, the primary loop scheme which is used for most residential areas in the United States will see mostly two hour outages (one every ten years). On the other hand, a spot network, which many utilities use in downtown areas and some industrial use for important processes, will see only two outages (one 12 hour and one 6 hour) for every 1000 years.

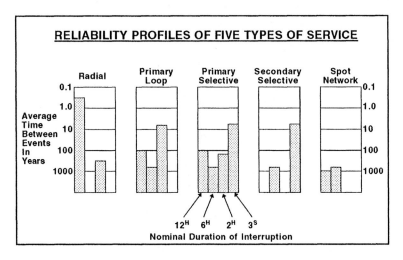

Figure 6-13. Reliability Profiles

Automation

It is an interesting paradox that when utilities are asked what is the most important advantage of distribution automation they answer reliability, while they also generally agree that their customers will not pay more for better reliability. Even though the inconvenience to the average customer during an outage is generally small, the cost to the utility can occasionally be quite high. Interruption cost assessment is a complex and often subjective task. Many of the effects resulting directly from interruptions are relatively easy to assign a dollar value. Other direct impacts such as reduction of manpower efficiency, fear, injury, and loss of life are difficult to quantify. Indirect effects such as civil disorder during blackouts or businesses moving to areas with higher reliability tend to be even more difficult to predict and evaluate. Many utilities do not feel than an outage costs them any significant amount of money.

A review of the literature was undertaken by Roy Billinton of the University of Saskatchewan and presented in his paper "Interruption Cost Methodology and Results - A Canadian Residential Survey". A plot summarizing the results is shown below.

255

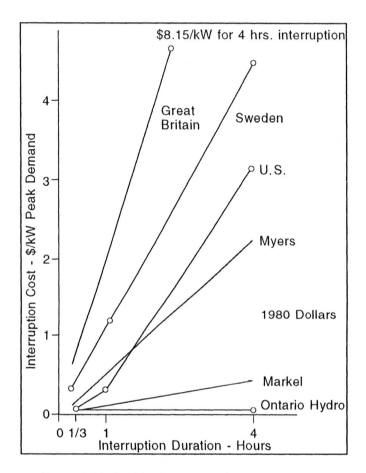

Figure 6-14. Residential Interruption Cost Estimates

Using these curves, and if we assume that a typical feeder outage lasts approximately 2 hours, we might conclude that the average interruption cost per kW is approximately $1.50. Assuming the typical distribution feeder shown below, we can begin to assess the cost per year of outages.

If we assume that this 13.8 kV feeder is approximately 7 miles long, and that the connected load is at least twice the diversified peak or approximately 20 MVA, we can begin to evaluate the value of increased reliability. Faults per year on the feeder would be calculated by multiplying the failure rate of .15 faults/mile/year by the length.

Consequently, the average number of faults per year on a feeder would be about 1. If we further assume that automation allows the fault to be isolated to 25% of the feeder main (assuming 3 sectionalizing points), a 75% reduction in outages can be recognized.

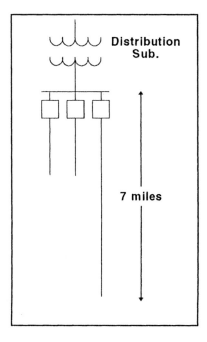

Figure 6-15. Typical Distribution Feeder Substation

The total savings per year per feeder based on the previous assumptions would then be .75 x 20,000 x $1.50 or $22,500 per feeder, per year. This number is apparently reflective of some utilities' assessments of liability and indicates the strong possibility that many of these utilities might justify automation on the basis of reliability only.

If, on the other hand, a utility is concerned only with the <u>lost revenue</u> costs because it believes that is its only cost, then the value of increased reliability due to automation might be calculated as follows:

Assume:

- Load factor = .5
- 1 fault/feeder/year
- 75% reduction in outage time due to automation [average duration = 2 hours]
- 10 cents per kWhr

Average MVA per feeder = 10 x .5 = 5.0 MVA/feeder

Savings = 5000 KVA x $.10/KWHR x 2 x .75 = $750.00/year/feeder

A savings of only $750.00 per year could be considered low and doesn't even take into account the fact that some of the projected revenue would not have been lost anyway (e.g., electric heat). The actual increased revenue is probably reduced to less than half or about $375/feeder/year.

System Modifications

There are several components that have successfully been used over the years to increase distribution reliability. A brief description of some of them is as follows:

Arresters. The purpose of lightning arresters on distribution systems is generally to protect equipment against failure due to lightning strokes (switching surges are usually too low to be a problem on a distribution system). Arresters can also be used to protect the distribution line from flashovers, a practice called "line protection". This practice can have a major effect in reducing the number of momentaries and consequently increasing power quality. Many utilities, in areas of high lightning activity use line protection. The problem with this concept is cost. In order for line protection to be very effective, arresters must be placed on all three phases of every pole or every other pole. Putting an arrester only on the top phase is not effective although some utilities subscribe to this practice. Table 6-4 illustrates the effectiveness of line protection practices.

Table 6-4. Line Comparison				
Design	Protection	Ground Resistance (ohms)	Percentage Flashovers	Flashovers per 100 miles a year
Crossarm	None	100	100.0	13.73
	Arresters on All Phases	100	2.4	0.33
	Arresters on Every Other Pole	25	28.5	3.91
		100	24.8	3.40
		250	27.5	3.78
	Shield Wire	25	11.1	1.52
		100	39.4	5.41
		250	61.3	8.42

Fuses. All reliability studies have shown conclusively that laterals should be fused. The generally accepted rule of thumb is that fuses larger than a 25K or a 15T should be used to avoid excessive nuisance fuse blowings. Nuisance blowing of transformer fuses can be the result of lightning or multiple inrush due to reclosing.

Switches. The addition of switches to a distribution circuit does, in general, increase reliability by decreasing the duration of the outage of many to the customers on the feeder. It is interesting to note, however, that the effectiveness of a switch is very much dependent on the length of the section being switched, switching practices, switch failure rate, etc. It has been found in some studies that the effect of a switch on system reliability can be very marginal and even have a negative effect. Table 6-5, shown below, illustrates the effect that sectionalizing has on a 34.5 kV, 19 mile feeder. As can be seen, without any sectionalizing the average customer minutes outage is approximately 51 minutes a year. With one switch (two sections), the average customer minutes outage is reduced to about 38 minutes or about 25%. To reduce the outage time another 25% will take an additional 3 switches.

Table 6-5. Effect of Sectionalizing

THE FEEDER LENGTH IS: 19
THE AVERAGE CUSTOMERS MINUTES OUTAGE FOR 12 KV IS: 17.5959
THE AVERAGE CUSTOMERS MINUTES OUTAGE FOR 35 KV IS: 51.3

35 KV SYSTEM OUTAGE IMPROVEMENT

NO. SECTIONS	OUTAGE AVG.
2	38.475
3	34.2
4	32.0625
5	30.78
6	29.925
7	29.31428
8	28.85625
9	28.5
10	28.215
11	27.98182
12	27.7875
13	27.62308
14	27.48214
15	27.36
16	27.25312
17	27.15882
18	27.075
19	27
20	26.9325

Automatic vs. Manual. The advantages of having automatic switching as opposed to manual switching is also dependent on a number of factors including crew size, crew procedures, number of sections, feeder length, etc. Figure 6-16, shown below, shows the effect of automatic switching for a 10 mile line. As can be seen, large numbers of automatic switches can prove quite effective (if the reliability of the switch and communication schemes are neglected).

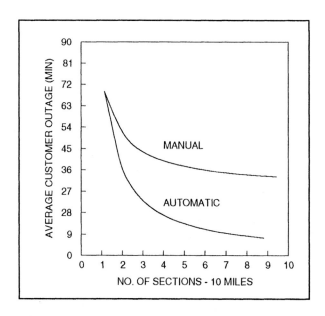

*Figure 6-16. Average Customer Minutes
Outage vs. Number of Sections - 10 Mile Line*

Underground Reliability

The reliability of an underground system is generally felt to be better than its overhead counterpart. Most distribution engineers would agree that while the frequency of failure is less for underground than overhead because of the lack of influence of most of the elements like ice, wind, cars, etc., the duration of an outage can be longer since the failure tends to be more difficult to locate and repair.

The following is a computer analysis of an underground system being served by various voltage levels. As will be seen, the reliability of an underground system is highly dependent on the system configuration, voltage (size) and the failure rates associated with the components.

Failure Rates. Equipment failure rates, as discussed earlier, are extremely important in any reliability study. When evaluating the effect that primary system voltage level has on failure rate the data becomes somewhat more difficult to interpret. While many utilities have a general "feeling" that the failure rate of higher voltage underground equipment

261

may be somewhat higher, few have quantified the data. Several utilities, who have had particular problems involving customer reliability have indeed quantified the failure rates at their various voltage levels and have shown failure rates to be considerably different than those shown in Tables 6-2 and 6-3. Several of these have also shown underground equipment failure rates to be many times higher at the higher voltage levels (25 kV or 35 kV) than at 15 kV. The baseline failure rates used for this analysis are shown in Table 6-6.

Table 6-6	
Circuit Element	**Failure Rate (per year)**
A. Primary underground	.07
cable (per mi.)	.006
B. Elbow disconnect	.001
C. Splice	.003
D. Transformer	.05
E. Switch	.005

Load Area Optimization. In developing reliability indices for different voltage levels, it is necessary to know the load geometry of the systems to be studied. No utility feeder main or lateral is the same and certainly great differences exist between utilities from the standpoint of voltage levels and loading which greatly affect the areas that can be covered by a given main/lateral design.

When comparing feeder areas operating at different voltage levels, it is necessary to treat the highly complex subject of distribution system planning in a generalized manner by making certain simplifying assumptions. In such a generalized study, the feeder load area is assumed to have uniform load density and a regular geometric shape, such as a rectangle or a triangle. While the results of generalized studies are not always directly applicable to a specific problem, they serve to illustrate the fundamental relations between percent voltage drop, losses, load density, circuit voltage, etc. An example of an area analysis for 3 different voltage levels is shown in Figure 6-17. As can be seen, the optimal way to increase system load at the higher voltages is to increase the length of both the feeder main and the laterals.

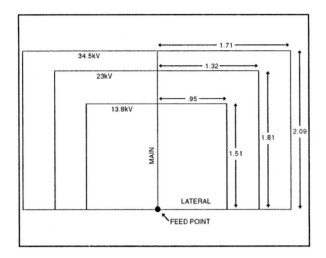

*Figure 6-17. Load Area Optimization at
Various Voltage Levels*

Computer Reliability Analysis. Reliability analysis of the optimized load areas for different voltage levels was performed in this example using two computer programs. The first analyzed the feeder main and the second the feeder lateral. Inputs to both are similar in that restoration procedures and times, protective configuration, line length, manual or automatic operation, etc. must be specified.

The three measures of reliability used by these programs was:

1. Annual interruption time per average customer,
2. Interruption time to last customer restored to service,
3. Annual number of interruptions per average customer.

Figure 6-18 represents the main circuit configuration used in the first program. It contains N - 1 automatic interrupters which divides the main into N equal-length sections. Each section contains M -1 manual sectionalizing points which divide the section into M is equal to or greater than 2. A circuit breaker is located at the substation where the main circuit goes underground. At the remote end of the main a normally open interrupter is located at an underground tie point to an adjacent similar circuit. The tie may be operated either automatically or manually. Single-phase underground laterals are spaced at regular

263

intervals along the entire length of the main. These laterals are connected to the main at the manual sectionalizing points in a manner which permits opening the main circuit on either side of the lateral connection.

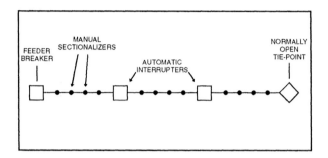

Figure 6-18. Main Current

Figure 6-19 shows the lateral circuit configuration used by the second program. Although it is not now common practice to use automatic interrupters out on the lateral, the program has provisions for m automatic interrupters, including the one at the connection point to the main.

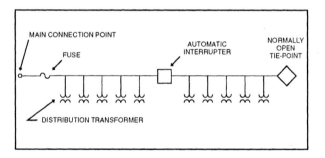

Figure 6-19. Lateral Circuit

For the purposes of this example, all reliability calculations were made assuming an underground system, use of fault indicators and fuses at all lateral taps. Both programs use "work functions" or restoration procedures, that can be tailored to a particular utility's experience and practice. A sample output showing reliability indices of feeder mains using standard failure rates is shown below. For example, if an optimized 23KV system had a feeder length of 1.81 miles, the system could expect an average customer minutes outage time for the main of about 12.75 minutes, a time to last customer restoration of 172 minutes and .181 interruptions per year.

Inputs and outputs for lateral reliability are somewhat similar to those performed on the feeder main. However, since calculations performed here had no intermediate protective devices in the lateral, the lateral reliability is simply a function of length. For example, in Figure 6-20 shown below, if a utility had a lateral length of 1.32 (as in the case of 23KV), the average customer minutes outage would be approximately 15.3 minutes.

```
AVAILABILITY CALCULATION FOR 23 KV

              MAIN FEEDER WITH AUTOMATIC TIE

                      WORK FUNCTIONS
TIME TO ARRIVE AT DEVICE (A):?  20
TIME TO CALL REMOTE LOCATION (B):?  20
TRUCK IN/OUT TIME (C):?  10
TIME TO OPERATE SWITCH (D):?  5
RESTORE TIME (LOCAL) (E):?  10
TIME TO READ FCI (F):?  10
TRAVEL RATE mi./min. (G):?  .25
              TERMINOLOGY

NUMBER OF CUSTOMERS PER UNIT AREA (K):?  800
MAIN CIRCUIT LENGTH (X):?  1.81
LATERAL CIRCUIT LENGTH (Y):?  1.32
OUTAGES PER UNIT LENGTH (J):?  .1
NUMBER OF SECTIONS (N):?  1
CONTROL CONSTANT (S):?  0
```

SECTIONS	AVERAGE INT. TIME	TIME TO LAST CUSTOMER	NUMBER OF INTERRUPTIONS
1	12.75	171.66	.181
2	6.38	168.33	.091
3	4.23	161.95	.060

Figure 6-20. Main Feeder Reliability Computer Output

```
LATERAL RELIABILITY PROGRAM FOR UNDERGROUND DISTRIBUTION SYSTEMS
                ***** WORK FUNCTIONS *****
        CASE NUMBER:?  1
        TIME TO ARRIVE AT LATERAL (A):? 20
        TRUCK IN/OUT TIME (C):? 10
        TIME TO OPERATE SWITCH (D):? 5
        RESTORE TIME (LOCAL) (E):? 10
        TIME TO READ FCI (F):? 10
        TRAVEL RATE mi./min. (G):? .25
        TIME TO OPERATE CONNECTOR (H):? 10
        NUMBER OF CUSTOMERS PER UNIT AREA (K):? 800
        OUTAGES PER UNIT LENGTH (L):? .16
        NUMBER OF SECTIONS (N):? 1
        NUMBER OF LAT. PER SEC. POINT (R):? 2
        CONTROL CONSTANT FOR TIE (0 for automatic) (S):? 0
        CONTROL CONSTANT (0 for radial) (V):? 1
        DISTANCE BETWEEN LATERALS (Z):? .08
        LATERAL LENGTH (Y):? .1

    LENGTH        AVERAGE CUST.       TIME TO          NUMBER
      OF             MINUTES        RESTORE LAST         OF
    LATERAL          OUTAGE           CUSTOMER       INTERRUPTIONS

     .1             1.28144           110.6            .0192
     .2             2.39648           111.2            .0352
     .3             3.52112           111.8            .0512
     .4             4.65536           112.4            .0672
     .5             5.7992            113              .0832
     .6             6.952639          113.6            .0992
     .7000001       8.115679          114.2            .1152
     .8000001       9.288321          114.8            .1312
     .9000001       10.47056          115.4            .1472
    1               11.6624           116              .1632
    1.1             12.86384          116.6            .1792
    1.2             14.07488          117.2            .1952
    1.3             15.29552          117.8            .2112
    1.4             16.52576          118.4            .2272
    1.5             17.7656           119              .2432
    1.6             19.01504          119.6            .2592
    1.7             20.27408          120.2            .2752001
    1.8             21.54272          120.8            .2912001
    1.9             22.82097          121.4            .3072001
    2               24.10881          122              .3232
    2.1             25.40624          122.6            .3392
    2.2             26.71328          123.2            .3552
    2.3             28.02992          123.8            .3712
    2.4             29.35616          124.4            .3872
    2.5             30.692            125              .4032
    2.6             32.03744          125.6            .4192
    2.7             33.39247          126.2            .4351999
    2.8             34.75712          126.8            .4511999
    2.9             36.13135          127.4            .4671999
    2.999999        37.5152           128              .4831999
```

Figure 6-21. Lateral Reliability Computer Output

Utilizing the optimized system configuration for 13.8, 23, and 34.5 kV systems, both programs were run and the systems total outage time (main and lateral) computed. A summary of the outage times for the various voltage levels is shown above. As can be seen in Figure 6-22, the inherent increase in customer minutes outage time by going from a 13.8 kV system to a 34.5 kV system is approximately 62%. However, the customer minutes outage time of the 34.5 kV system is still considered very low when compared to the overhead systems.

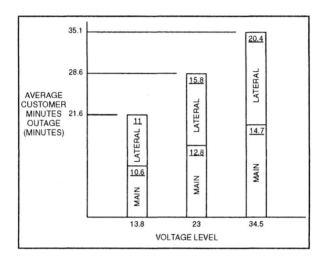

Figure 6-22. Voltage vs. Outage Time

Some utilities have indicated that their failure rates with higher voltage components are even higher than those shown in Table 6-4. If the assumption is made that the failure rate of 23 kV equipment is twice that of 13.8 kV and that 34.5 kV is three times the failure rate of 13.8 kV equipment then a graph of the outage times vs. voltage (see Figure 6-23) would be considerably more dramatic. As can be seen, the average customer minutes outage now increases to 105.5 minutes or almost 300%. This of course means that a utility, unsuspectingly facing this high failure rate, may have to take drastic corrective action.

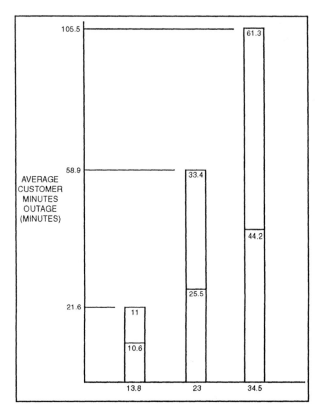

Figure 6-23. Voltage vs. Outage for Higher Failure Rates

REFERENCE

1.   G. Wacker, E. Wojczynski, and R. Billinton, "Interruption Cost Methodology and Results - A Canadian Residential Survey", *IEEE Trans. on Power Apparatus and Systems*, Oct. 1983, pp. 3385-3392.

QUESTIONS

1.   Average Customer Minutes Outage is defined by which indices?

2.   Customers seeing an average frequency of interruption of 2 per year should consider themselves fortunate. (true or false?)

3.   The leading cause of faults is _____.

4.   Transformers have a high failure rate. (true or false?)

5.   "Blocking the instantaneous" improves power quality _____ (always, sometimes, never).

6.   Higher voltage systems are usually less reliable. Explain.

7.   Primary selective systems eliminate the "blinking clock syndrome". (true or false?)

8.   Laterals should only be fused if coordination with the feeder breaker (or recloser) is possible. (true or false?)

9.   One way to decrease the number of momentary outages is to ____.

10. Adding switches is a good way to increase reliability. (true or false?)

11. "Blocking the instantaneous" hurts reliability. Explain.

12. What are the advantages and disadvantages of 34.5 kV?

13. Explain the advantages and disadvantages of underground distribution.

7

POWER QUALITY FUNDAMENTALS

INTRODUCTION

Good power quality is not easy to define because its measure depends upon the needs of the equipment which it is supplying; what is good power quality for a refrigerator motor may not be good enough for a personal computer. For example, a short (momentary) outage would not noticeably affect equipment such as motors or lights, but could cause a major nuisance to digital clocks or VCRs.

DEFINITIONS

Probably no other area in power quality has caused more confusion than that of definitions. Definitions have varied from author to author. The biggest difference between definitions seems to exist between electronics engineers and power engineers. The definitions we will follow in this chapter are shown in Figure 7-1 and defined as follows:

Outages (Interruptions). An outage is a complete loss of voltage usually lasting from as short as 30 cycles up to several hours, or in some cases even days. Outages are usually caused by the fault induced operation of circuit breakers or fuses. Some of these interruptions might be classified as permanent, while others might be classified as temporary (momentaries).

Surges (Lightning or Switching Surges). A surge is a transient voltage or current which can have extremely short duration and high magnitude. Typically, surges are caused by switching operations or by lightning. Surges can be generated by customers through the switching of their own loads, or they may be caused by utility switching operations (capacitors, breakers, etc.). Surges have always existed on power systems, but it is only in recent years, with the advent of extensive VCR and PC use, that they have received attention.

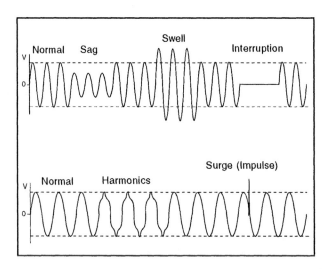

Figure 7-1. Typical Voltage Disturbances

Undervoltage (Voltage Drop). A customer who experiences a long duration (several seconds or longer) service or utilization voltage which is less than the proper nominal operating low voltage limit can be considered to be experiencing an undervoltage situation. (The ANSI Range [A] service voltage and utilization low voltage limits are 114 volts and 110 volts respectively.) Two examples of factors which may cause such a condition are: 1) overloaded or poor house wiring, and 2) poor connections and/or voltage drop on the utility system.

Harmonics. These are the nonfundamental frequency components of a distorted 60 Hz waveform. They have frequencies which are integral multiples of the 60 Hz fundamental frequency. Harmonics are not generally produced by the utility, but rather by the customer's equipment. For example, a large nonlinear industrial load may produce harmonics which, if they are of sufficient magnitude, can travel back through the power system and affect other customers.

Voltage Sags. A momentary voltage dip that lasts for a few seconds or less is classified as a voltage sag. Voltage sags may be caused by faults

271

on the transmission or distribution system, or by the switching of loads with large amounts of initial starting/inrush current (motors, transformers, large dc power supplies). Voltage sags may be sufficiently severe, especially in the case of faults, to cause sensitive loads (computers, VCRs, clocks, etc.) to reset.

Voltage Swell. When a fault occurs on one phase of a 3-phase, 4-wire system, the other two phases rise in voltage relative to ground (about 20%). This steady state rise in voltage is referred to as a <u>swell</u>. Voltage swells usually have durations of several seconds or less, but can last as long as a minute or so.

Overvoltage. An overvoltage is classified as any steady state voltage (several seconds or longer) which is above the ANSI Standards' upper service voltage limit of 126 volts at the customer's meter. Overvoltages usually occur as a result of improper regulation practices (misadjustments of regulators and capacitors).

VOLTAGE QUALITY

Voltage Drop

Almost all equipment connected to a utility system is designed to be used at a certain definite voltage. It is not practical, however, to serve every customer on a power distribution system with a constant voltage corresponding exactly to the nameplate voltage. This is because voltage drop exists in each part of the power system from the generator to the customer's meter. There is also considerable voltage drop in the consumer's internal wiring. Since voltage drop is proportional to the magnitude of the load current flowing through the entire power system impedance, the customer who is electrically farthest from the substation will receive the lowest voltage (see Figure 7-2).

Since all customers have essentially the same utilization devices, it is necessary to provide each of them with nearly the same utilization voltages. A compromise is needed, however, between the voltage range which the utility must supply, and the voltage range within which equipment will operate satisfactorily. If the limits of voltage provided by the power company are too broad, the cost of appliances and other utilization equipment, especially computers, will be high because they will have to be designed to operate satisfactorily within these limits. On the other hand, if the voltage limits required for satisfactory operation of the utilization equipment are too narrow, the cost of providing power within these limits will be excessively high.

Over the years, electric utility companies and equipment manufacturers have cooperated in establishing standards for operating voltage limits which have proved to result in satisfactory operation without excessive demands upon the design of the power system or on the utilization of equipment.

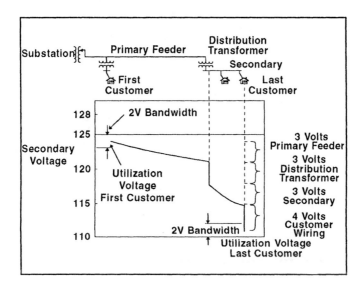

Figure 7-2. Voltage Profile - Residential Feeder

The American National Standards Institute (ANSI) has established Standard C84-1, "Voltage Ratings for Electric Power Systems and Equipment," which was formulated by both utilities and manufacturers, and its recommendations are followed by both. Figure 7-3 illustrates the standards for service (meter) and utilization (load) voltages. The utility must only meet the "service" requirements because only the customer has control over the voltage drop in his circuit. The Range A values are defined as the limits within which systems shall be designed and operated so that most of the service voltages fall within them (114-126). Voltage variances outside these limits are to be infrequent. Voltage Range B levels are allowable provided they occur infrequently and are of limited duration (110-127). When they occur, corrective measures are to be undertaken within a reasonable amount of time to provide voltage within Range A limits.

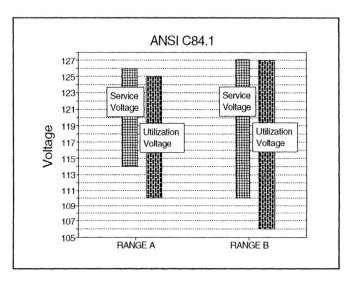

Figure 7-3. Voltage Ranges Specified By ANSI Standard C84.1-1984

The utility has some control over voltage to the customer's load. This control, however, is not instantaneous and does not ensure a constant voltage, but merely maintains the voltage between some desired limits. While standard voltage regulating practices may be effective for gradual voltage changes, they have little effect on voltage flicker.

Typical voltage profiles for heavy and light loads are shown in Figure 7-4. These profiles show, in a conceptual way, the voltage drops or rises that may occur on the overall system between the generators and the loads to be served. During any one day, there is a definite difference in voltage levels between "heavy load" and "light load" periods. Under heavy load, the power flows (and thus the voltage drops) are larger. Under light load, there will be less voltage drop. Notice that voltages are corrected throughout the power system, so that excessive drops do not occur. For short distribution feeders, it is entirely likely that voltage standards can be met by using only the voltage control equipment in the substation, such as LTC's (Load Tap Changing) or regulators. However, for long feeders or feeders with extremely heavy loads, it may be necessary to augment the substation equipment with voltage regulators located out on the feeder.

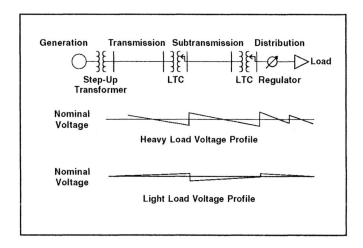

Figure 7-4. Voltage Profile Across System During Heavy and Light Load

In addition to LTC transformers and regulators, many utilities employ shunt capacitor banks to help control the feeder voltage profile. Because capacitors supply reactive power (VARs) to a circuit, they can effectively reduce and even cancel out the VAR requirements of inductive loads, such as motors. Placed strategically on the feeder, shunt capacitors will not only improve voltage profiles, but will also reduce the resistive losses associated with the primary feeder and laterals.

Figure 7-5 illustrates the improvement in voltage obtained by connecting a three-phase capacitor bank in the primary feeder. Without the capacitor (solid line), the service voltages range from 105% (first customer) to 90% of nominal (108 volts at the last customer). With the capacitor connected from the feeder to ground, the voltage profile will be adjusted as shown by the dotted line. Voltages along the entire feeder are now within plus or minus 5 percent of nominal.

Sometimes, especially during light load conditions, regulators or capacitors can overcompensate and create an overvoltage lasting from a few seconds to many hours (see Figure 7-6).

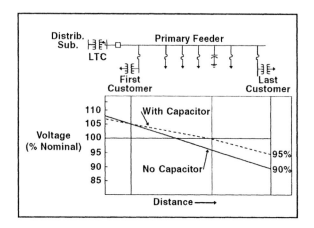

Figure 7-5. Voltage Profile With Capacitor

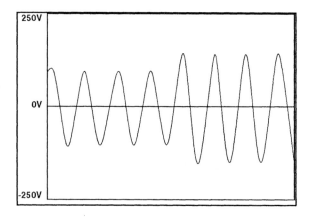

Figure 7-6. Overvoltage Condition

Figure 7-7 shows the effect of putting a large bank of capacitors on a system at light load. As can be seen, the utility system voltage, which is usually limited to 127 volts (126 volts at the customer's meter), now approaches 130 volts and is above the ranges suggested by ANSI C84.1.

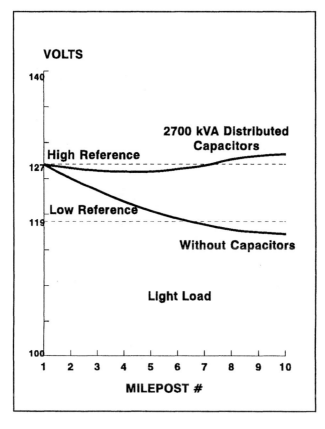

Figure 7-7. Effect of 2700 kVAR of Distributed Capacitors

Flicker

Many voltage problems associated with computers, however, are not necessarily related to simply high or low voltage, but rather to rapid changes in voltage. This is called "voltage flicker." Utility voltage regulation equipment, which will be discussed later, will not compensate for instantaneous voltage fluctuations caused by the sudden application of low power factor loads such as motors. To illustrate, Figure 7-8 shows a power system supplying an arc furnace and some domestic loads where voltage flicker might be a problem.

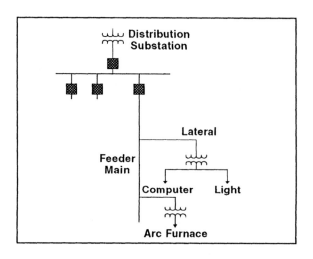

Figure 7-8. Simplified Circuit Illustrating the Flicker Problem

Whereas the domestic load currents are practically constant, the arc furnace load undergoes rapid changes and generates a variable voltage dip along the distribution feeder. The voltage of the feeder primary becomes variable, and disturbances may be noticed by some of the domestic loads attached to it.

Figure 7-9 shows a curve of the maximum permissible voltage flicker in incandescent lighting before it becomes objectionable to the customer (borderline of irritation). If a user, for example, is in the vicinity of a heat pump or air conditioner which fluctuates approximately 6 times per hour, and notices that his incandescent lamps are flickering, he can assume, using this curve, that the voltage change is on the order of about 4% (or more). In many cases, the computer may be far more sensitive to voltage change than the human eye and consequently experiences problems the human eye cannot detect.

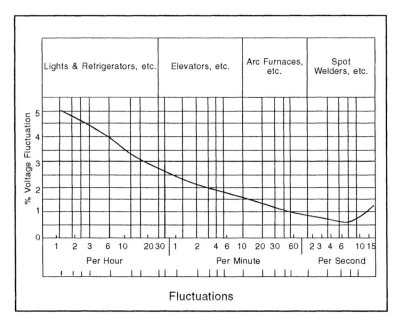

Figure 7-9. Voltage Flicker - Border Line of Irritation Curve

Voltage Sags

A voltage sag is a short term decrease in steady state voltage. These sags, sometimes lasting many seconds, can either have very little effect on sensitive loads if the drop in voltage is not more than 10% or 20%, or they can have a major effect (similar to an outage) if the voltage sag is larger (e.g., 50%). Voltage sags in an industrial facility are generally caused by large loads switching on. This can also happen on a utility, but more often, severe sags are caused by faults on the system. Figure 7-10 illustrates a severe sag caused by a transformer fault. Most severe sags on a utility system are the result of faults on the same feeder, at the substation, or even on an adjacent feeder, and usually, they are relatively easy to identify. A common mistake is to believe that a power line conditioner will protect a sensitive load from sags. While this is true for some sags, it is not usually the case for severe sags where the PLC (power line conditioner) cannot fully compensate for the voltage dip and the undervoltage relay of the PLC drops out.

279

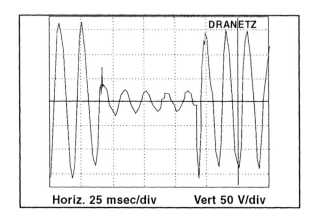

Horiz. 25 msec/div   Vert 50 V/div

Figure 7-10. Sag Caused by Transformer Failure

Figure 7-11 illustrates another interesting situation. As shown, faults on the transmission and subtransmission system can affect customer voltages over 50 miles away. Sometimes these voltage sags can be severe and cause equipment, such as PLCs or computers, to trip out. When a customer calls to complain of poor power quality, the real problem for the utility is that they may have no local record of a system disturbance because the fault was not in the distribution system, and possibly not even in the same district.

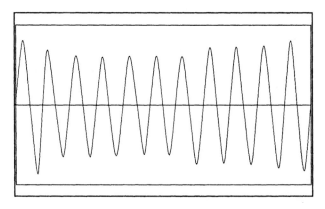

*Figure 7-11. Power Quality Disturbance: Voltage
Sag due to Subtransmission Fault*

Voltage Swells

The occurrence of ground faults on 3-phase systems cause the voltages in the unfaulted phases to rise with respect to ground (electronics engineers refer to these as "surges") (see Figure 7-12). This voltage increase can be as much as approximately 30% for a 4-wire, multigrounded system and over 70% for a 3-wire system. The duration of this overvoltage, or "swell," is dependent on the system protection, which can take as long as minutes or as little as a half cycle.

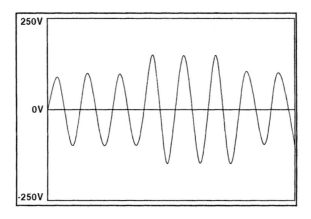

Figure 7-12. Voltage Swell

The magnitude of overvoltages during ground faults are particularly important to the proper sizing and operation of surge arresters. The standards in this area are hence developed by the Surge Protective Device Committee of IEEE. The maximum magnitude of overvoltages (swells) occurring for various line-to-ground faults for different distribution system designs (per IEEE C62.92-1991) is shown below:

System	Overvoltage Magnitude
Ungrounded	$1.82 \times E_{LG}$
Four-wire multigrounded (spacer cable)	$1.5 \times E_{LG}$
Three- or four-wire unigrounded (open wire)	$1.4 \times E_{LG}$
Four-wire multigrounded (open wire-gapped)	$1.25 \times E_{LG}$
Four-wire multigrounded (open wire-MOV)	$1.35 \times E_{LG}{}^{*}$

E_{LG} = Nominal line-to-ground voltage of system

*Because the metal-oxide varistor (MOV) arrester is more sensitive to poor grounding, poor regulation, and the reduced saturation sometimes found in newer transformers, many utilities are using a more conservative 1.35 factor

Harmonics

In the ideal power system, the voltage supplied to customer equipment and the resulting load current are perfect sine waves. However, conditions are never ideal in practice, so these waveforms are often distorted. This deviation from perfect sinusoids is usually expressed in terms of harmonics distortion of the voltage and current waveforms. Harmonics distortion problems are not new to utilities. In fact, such distortion was observed by utility operating personnel as early as the 1920's. Typically, the distortion was caused by nonlinear loads connected to utility distribution systems. For example, an arc furnace is nonlinear since it will draw a nonsinusoidal current (rich in harmonics) when a sinusoidal voltage is applied (see Figure 7-13A). The distorted load current then causes distorted bus voltages to appear throughout the system (see Figure 7-13B). In the past, such harmonic sources were not very widely used, and harmonics were often effectively mitigated through the use of the grounded wye-delta transformer connection.

Today, however, additional methods for dealing with harmonics are necessary due to the influence of the following developments:

1. The recent proliferation in the use of static power converters,
2. Added network resonances, and
3. Power system equipment and loads which are more sensitive to harmonics.

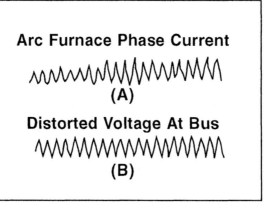

Arc Furnace Phase Current

(A)

Distorted Voltage At Bus

(B)

Figure 7-13. Harmonic Distortion

There are now two criteria that are used to evaluate harmonic distortion. The first is a limitation in the harmonic current that a user can transmit into the utility system. Table 7-1 lists the harmonic current limits based on the size of the user's harmonic loads with respect to the size of the power system to which he is connected. The ratio of I_{sc}/I_L is the short circuit current available at the point of common coupling (PCC) to the nominal fundamental load current. Thus, as the size of the user load decreases with respect to the size of the system, the larger is the percentage of harmonic current the user is allowed to inject into the utility system. This protects other users on the same feeder as well as the utility, which is required to furnish a certain quality of power to its customers.

The second criterion specifies the quality of the voltage which the utility must furnish the user. Table 7-2 lists the voltage distortion that is acceptable from a utility by a user. This table is similar to the one in the present edition of IEEE 519. Since the utility is the PCC between users, it has the responsibility of monitoring the harmonic current each user injects into its system, and of ensuring that this current does not cause its voltage to reach distortion levels higher than those listed in Table 7-2. The values of Table 7-2 are low enough to ensure that equipment will operate correctly.

Table 7-1. Harmonic Current Limits (in %) from IEEE Std. 519						
I_{SC}/I_{LOAD}	HARMONIC ORDER					Total Harmonic Distortion
	<11	11-16	17-22	23-24	>35	
< 20	4.0	2.0	1.5	0.6	0.3	5.0
20-50	7.0	3.5	2.5	1.0	0.5	8.0
50-100	10.0	4.5	4.0	1.5	0.7	12.0
100-1000	12.0	5.5	5.0	2.0	1.0	15.0
> 1000	15.0	7.0	6.0	2.5	1.4	20.0

Where I_{SC} = Maximum short circuit current at point of common coupling.
And I_L = Maximum demand load current (fundamental frequency) at point of common coupling.
TDD = Total demand distortion (RSS) in % of maximum demand

$$TDD = \sum_{h=2}^{H} \left(\frac{I_h^2}{I_{L\ demand}} \right)^{1/2} x\ 100\% = Total\ harmonic\ distortion$$

Table 7-2. Maximum Voltage Distortion per IEEE Std. 519			
Maximum Distortion (in %)	SYSTEM VOLTAGE		
	Below 69 kV	69 - 138 kV	> 138 kV
Individual Harmonic	3.0	1.5	1.0
Total Harmonic	5.0	2.5	1.5

(For conditions lasting more than one hour. Shorter periods increase limit by 50%)

Harmonics are not normally produced by the power system itself but rather by the loads connected to the power system. Some of the more common sources of these harmonics are:

- Static power converters
- Overexcited transformers
- Fluorescent lights
- Solid state devices (computers, dimmer switches, variable speed drives).

For example, solid state switching devices may use diodes, thyristors, etc. The simple single-phase, full-wave rectifier, shown in Figure 7-14 illustrates the operation of this commonly used device. As can be seen, although the voltage input waveshape is a sine wave, the current drawn by the device is rich in harmonics.

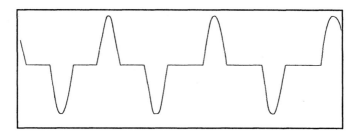

Figure 7-14. AC Drawn by Consumer Equipment

Likewise, a transformer operating at a voltage higher than its normal operating point may be pushed into its saturation region (see Figure 7-15) where large harmonic content is produced (see Figure 7-16).

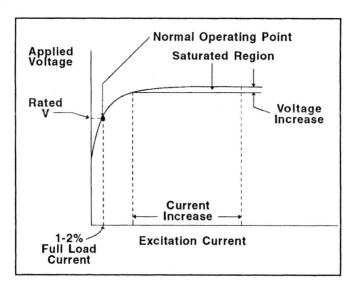

Figure 7-15. Transformer Excitation Current Plotted vs. Applied Voltage

285

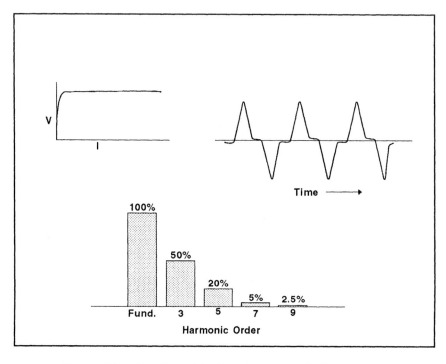

Figure 7-16. Nonsinusoidal Transformer Excitation Current
and its Harmonic Content

Other examples of wave distortion are shown below in Figure 7-17:

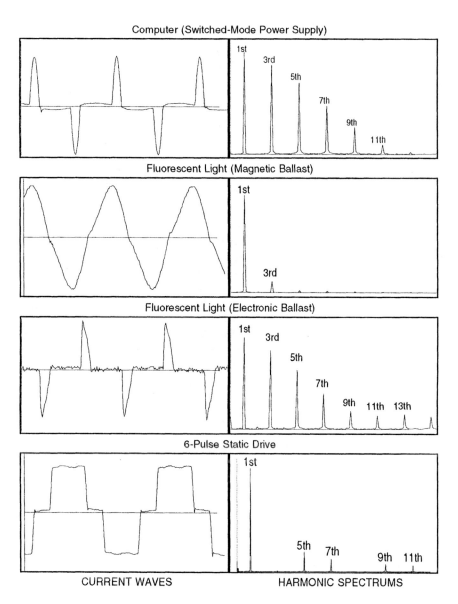

Figure 7-17. Example Waveforms From Several Common Sources.

Harmonic problems on utility systems are generally difficult to diagnose since the level of harmonic buildup may be gradual or a system change totally unrelated to a harmonic source may result in a new problem. The most common harmonic-caused problems found on the distribution system would normally manifest themselves as one of the following:

- Excessive capacitor fuse blowing
- Motor and transformer overheating
- Unexplained breaker tripping due to a ground fault relay
- Complaints of telephone interference.

The solution to these problems range anywhere from simply changing the location of something like a capacitor bank to having to replace a transformer or put in a filter. This is one area of distribution engineering where "an ounce of prevention may be worth a pound of cure", i.e., don't exceed IEEE 519 recommendations.

Example 1. Suppose a utility has a customer who wants to install a device, at 13.8 kV, where the continuous load is 100 amperes and the total demand distortion (TDD) is 7%. The location of his business is at a point on the feeder where the short circuit current is 1600 amperes. Should the utility allow this load? And if not, what can be done to accommodate this customer?

Solution. I_{sc}/I_L = 1600/100 = 16

IEEE 519 indicates (see Table 7-1) that if the TDD is 7%, the ratio of short circuit to load current must exceed 20. Since the actual ratio is only 16, the customer has two choices: relocate the business closer to the substation to increase the short circuit level to at least 2,000 amperes (very unrealistic) or put in filters to reduce the TDD to 5% or less.

Interruptions

Service outages can be caused by either planned or unplanned outages of equipment. When equipment outages are planned, such as to increase the transformer size, the supply to a small number of customers may be intentionally interrupted. More likely, however, service outages are caused by unplanned events (faults), such as lightning, wind, or ice. In the case of either planned or unplanned outages, the system is designed and the protection schemes are operated so that:

1.   The number of service outages is minimized;
2.   The duration of a service outage is minimized;
3.   The number of customers interrupted is minimized.

Faults on the utility system are classified as either temporary or permanent. A temporary fault may be due to a lightning stroke, an animal, wind, or other natural occurrence. When the fault occurs, the line must be de-energized to stop the flow of fault current and enough time must be allowed so that the faulted path can be de-ionized. To do this, the breaker in the substation will open immediately to clear the fault, and then automatically reclose after some time delay. This reclosing can occur several times in an effort to re-establish continuity of service for a temporary fault. Typical time delays between reclosures are shown in Figure 7-18.

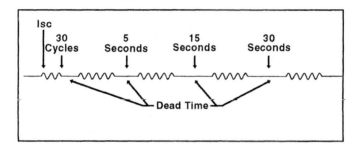

Figure 7-18. Current vs. Time

This opening and reclosing of the breaker is usually short enough that most customers may not even be aware that service had been interrupted. A computer or other sensitive load, however, will see a total system shutdown unless steps have been taken (uninterruptible power supply) to avoid this short outage.

A permanent fault will result in a service outage which may last several minutes to an hour or more. The average duration of a permanent fault per a recent survey was approximately 2 hours. During a permanent fault condition, the breaker will usually operate three or four times in an attempt to reestablish power before it locks open. A broken conductor,

289

a tree limb falling onto a conductor, or perhaps across two conductors, are all examples of permanent faults. In this case, the fault must first be located and repaired before service is restored to all customers.

Most conductor-related outages on overhead distribution lines are of a temporary nature (roughly 75%). By contrast, most faults on underground systems are permanent and take much longer to locate and repair. One consequence of having more sensitive loads is that the 75% of overhead faults which were once classified as temporary are now classified as causing permanent equipment outages.

The frequency and duration of faults is <u>to a limited degree</u> a function of the utility system design. There are some basic system designs used by utilities which are more reliable than others. Unfortunately, as the reliability goes up, so (usually) does the cost.

One major complication to reliability calculations is that, due to sensitive loads, what was once not considered an outage is now considered an outage. This is certainly the case in successful reclosing on a temporary fault.

This re-evaluation caused by more sensitive loads has changed some utilities philosophy with regard to "feeder selective relaying". Feeder selective relaying simply means that for the system shown in Figure 7-19, the fuse should blow for permanent faults and the breaker should operate faster than the fuse for temporary faults.

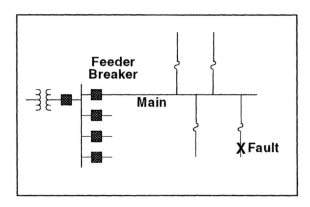

Figure 7-19. Feeder Selective Relaying

Some utilities that have considerable penetration of computer loads on these laterals prefer not to reclose at all, even for a temporary fault.

This is because the breaker operation affects the entire feeder, while the operation of the fuse affects only the lateral with the temporary fault. This does, however, create a permanent outage for those on the faulted lateral tap.

Voltage Surges

Voltage surges on a power system are the most common sort of power problems seen by the computer user. These transients can be the cause of lost data, false triggering, and equipment failure, and many of them are generated internally by the user himself. Other transients are the result of some occurrence on the utility primary, such as lightning and equipment switching.

Lightning transients to the low voltage system can occur from either direct strikes to the secondary circuit or strikes to the primary circuit where transient voltages pass through the distribution transformer. Although those transients which pass through the transformer are reduced in magnitude, they are not reduced by the turns ratio of the transformer (approximately 60 to 1) because the transformer windings appear, electrically, to be a capacitor. Lightning transient voltages on the utility primary are limited to equipment flashover levels (approximately 95 kV-300 kV for a 15 kV class system) and by arrester protective levels which are approximately 40 kV crest (see Figure 7-20).

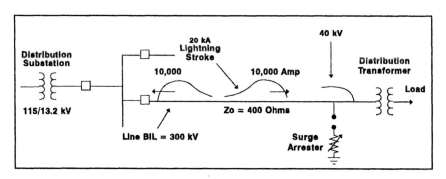

Figure 7-20. Lightning Induced Voltage Surge

SENSITIVE LOADS

Household Appliances

Some engineers have argued that this present emphasis on power quality is the direct result of the sensitivity of modern digital appliances. The "blinking clock syndrome" is certainly well known to all distribution engineers. The previous analog clocks were somewhat immune to the successful reclosing sequence which transpired during a temporary fault. A digital clock without battery backup, however, interprets a momentary interruption the same as it would a complete outage. Figure 7-21, shown below, illustrates the sensitivity of 3 of our most common household appliances. As can be seen, virtually none of these appliances could operate properly after even an instantaneous reclosure, since these take 20 cycles or more and most devices can't ride through even a cycle of no voltage. A sag, on the other hand, is another story. All three curves indicate that a voltage sag that does not reduce the voltage to less than 60% of nominal should pose no problem. Sags on the transmission system generally would not be expected to cause a sag of this magnitude.

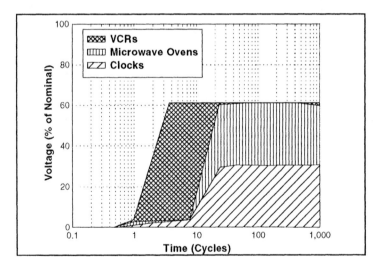

Figure 7-21. Malfunction Range Summary For All Equipment

Computers

Computer equipment is more sensitive to power quality than most other equipment. The CBEMA curve (Computer Business Equipment Manufacturers' Association) was created in order to define the transient and steady state limits within which the input voltage can vary without affecting the proper performance of or causing damage to the computer equipment, as shown in Figure 7-22.

Referring to the figure, it can be seen that a range from +6% to -13% of the nominal voltage will allow the computer to function properly. For shorter time spans, the voltage tolerances are larger. For example, voltage can fall to zero as long as it recovers with 0.5 cycles or 8.33 ms and it can drop to 30% below normal for up to 0.5 seconds. This is because stored energy is built into the computer's power supply filters and motors.

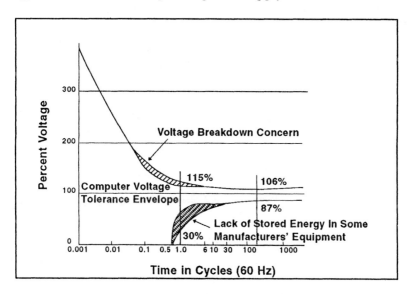

Figure 7-22. CBEMA Curve

Low Voltage Environment

Power quality problems at the customer level are the result of both utility created disturbances as well as those created by the customer himself. The residential customer influences their own power quality either by the design of their house wiring, their use of appliances, or the types of load used. Some of the more important findings concerning the environment below 600 volts are the following:

Swells. Typical overvoltages in the range of 5% above nominal would not be expected to have any serious effect even on sensitive loads. Swells, on the other hand, are severe steady state overvoltages on the unfaulted phases of a system experiencing a line-to-ground fault. These overvoltages can be expected to approach 1.25 per unit on the typical 4-wire multigrounded system. Figure 7-23 shows an actual case where the voltage measured in the home reached 150 volts. The duration of this overvoltage is a function of the speed of the protection system. Tests by one utility showed that swells of 22% lasting 120 cycles caused no damage or misoperation of typical sensitive loads. These same tests have not been performed repetitively or upon aged equipment so it is difficult to assess the overall effect of this condition.

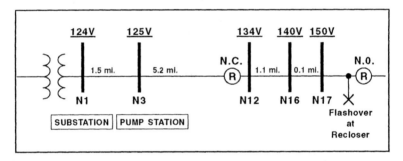

Figure 7-23. Overvoltages During Line-to-Ground Fault

Harmonics. Harmonics standards (IEEE 519) have been discussed earlier and indicate that the utility system will tolerate only very moderate amounts of current and voltage harmonic distortion. It should be noted that a household appliance, such as a stereo can generate significant amounts of current distortion (see Figure 7-24). Although the percent distortion for this device is high, the voltage distortion is low since the source is relatively stiff and the current consumption of such a device is low (see Figure 7-25).

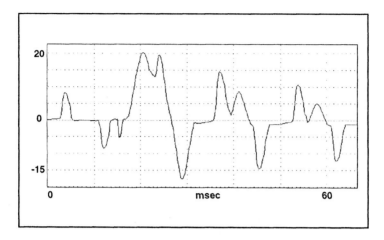

Figure 7-24. Current Measured During Stereo Operation.

Figure 7-24 shows the voltage waveshape measured at the same time as the current.

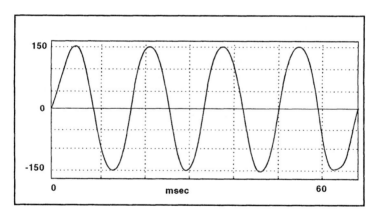

Figure 7-25. Voltage Measured During Stereo Operation.

Sags. Sags in the home caused by the utility are generally infrequent but quite severe. For very sensitive loads, or for severe sags, misoperation can occur. Sags caused by the home or industrial complex are more likely to be caused by motor startups and are generally not severe enough to

295

cause any problem. Tests with air conditioners indicate that normal air conditioning operation can result in significant sags. For example, in a test of a standard house window unit, the voltage sag measured at the meter during startup was only 3 volts while the voltage sag due to the drop in the house wiring was over 30 volts.

Surges. Studies have shown that surges found in the home are almost all created by conditions within the home itself. These voltages can be as high as 2000 volts but are more commonly in the range of 100 to 500 volts. They do not seem to cause any perceptible problems.

POWER LINE CONDITIONING

Introduction

Utility engineers throughout the country are finding that knowledge of just their power system is inadequate when working in the area of "power quality". More and more, utility customers, who are experiencing power related problems for sensitive loads, are depending on the utility for solutions, even when the only solution is on the customer's premises. For this reason, it is now important that distribution engineers have a general understanding of the options available to the customer and the limitations of those options.

Power line conditioners mean different things to different people. The term power line conditioner can be applied to any device that protects against one or more of the following power problems:

- Overvoltages and undervoltages
- Sags
- Outages
- Surges
- Harmonics

There are some power disturbances which exist all the time but do not cause any malfunction or component stress on computers or similar delicate electronic equipment. For example, there is always some harmonic distortion on the power line sinewave, and there are occasional transients. These distortions generally cause no harm.

The price range of power line conditioners can vary from $25 to well over $10,000, and one cannot always be sure what is being purchased for that price. Generally, complete specifications are not found on the product, and if an attribute is not described on the packaging of the device, it will more than likely not have that capability. A line conditioner should solve most power quality problems, and its specifications should cover each type of disturbance listed above. This section will briefly discuss the advantages and disadvantages of the more commonly applied devices.

Surge Suppression

Surge suppressors are devices which conduct electricity when the voltage exceeds a limit. The device does not really suppress the surge (like a filter) but rather diverts it (usually to ground). Figure 7-26 shows the effect of the surge suppressor known as an MOV (Metal Oxide Varistor).

The amount of energy these suppressors can absorb is directly related to their size. If the device is overstressed, it will self destruct; thus it is very important that the energy level of the transient for a particular location be known. For example, suppressors at the panelboard will be much larger than those generally found in wall receptacles or on electronic devices.

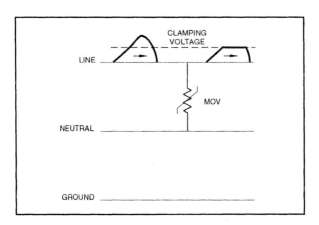

Figure 7-26. Normal Mode Surge Suppressor

Evaluating the performance of a surge suppressor is not an easy task. The configuration of components (line-to-ground, line-to-neutral, neutral-to-ground), as well as the quality and characteristics of the components are not easily determined. Marketing tends to disclose only what the devices will do and nothing about what they will not do.

The best surge suppression on the utility primary distribution system will not be adequate for the protection of a customer's sensitive equipment. Voltage transients coming from the utility system, as well as those (over 90%) generated internally, can only be mitigated on the secondary at the customer's premises (see Figure 7-27). This power conditioning is cheap and, in the author's opinion, mandatory for good performance.

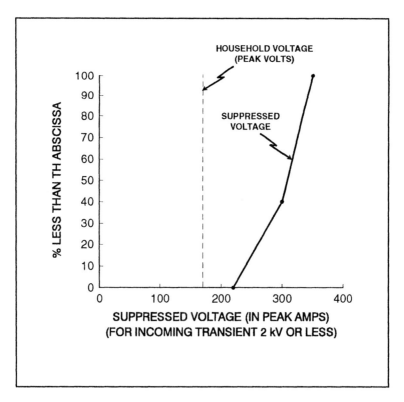

Figure 7-27. Summary of Surge Suppressor Data Average Value of Suppressed Transients

Constant Voltage Transformers (CVTs)

A ferroresonant transformer is essentially a transformer operating in a saturated mode. Ferroresonant transformers (see Figure 7-28) regulate voltage and, to a degree, can perform wave shaping.

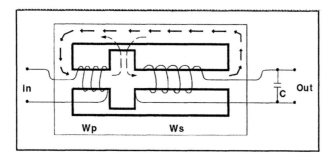

Figure 7-28. Elements of Ferroresonant Regulation Using a Transformer

While the CVT has many positive qualities which made it very popular for dc power supplies (IBM used nothing but ferroresonant dc power supplies in their computers for many years), they do have the following disadvantages:

a. The transformer leakage impedance limits the short circuit current, which is good. This impedance however, can become a severe limitation when it involves start-up currents of a motor and other electronic equipment.

b. If more than one device is on the transformer, start-up of one can cause the magnetic field to collapse. This can result in the misoperation of the other(s).

c. Since the circuit (essentially a tank circuit) stores energy, it has some ability to "ride through" certain minor disturbances. If a tank circuit loses power, then it must replace this power. Because of this, a one-cycle outage can look like a multicycle voltage sag, which may have a greater effect on the proper performance of the equipment.

d. Devices are heavy, noisy, and cause output transients which can greatly interact with computers.

Isolation Transformers

Figure 7-29 shows how an isolation transformer was used to break the ground loop in order to eliminate common-mode problems (this is now prohibited by Federal law). These types of problems are reduced because a common-mode pulse cannot develop a magnetic field in the core of the transformer; yet the common-mode, high frequency impulses can still pass through to the secondary because of the transformer coupling capacitance. Common-mode attenuation capabilities can be substantially improved in an isolation transformer by using electrostatic Faraday shielding around the windings. The shields effectively add bypass capacitance which routes high frequencies to ground.

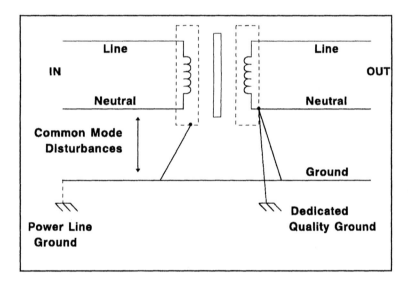

Figure 7-29. Isolation Transformer

Some of the drawbacks of isolation transformers are as follows:

*   Cannot remove any normal-mode signal or disturbance.
*   Shielding can form an LC circuit which is resonant at some high frequency.
*   Can cause severe power loss that causes clipping of the sinewave peaks that are above nominal rating.

- Can saturate during transients and distort the next few cycles.
- Saturation currents can result in nuisance fuse blowings.

Line Conditioners

While power line conditioning can, in most people's minds, refer to anything from a simple surge arrester to a complex UPS system, a power line conditioner generally brings to mind a device which provides several functions, including impulse attenuation, filtering, voltage regulation, and isolation. In truth, some line conditioners do provide these functions, and some do not ... so be careful. Two of the more common types of line conditioners are described as follows:

a. **Enhanced Isolation Transformer:** The enhanced isolation transformer, shown in Figure 7-30, uses the inductance of the chokes to attenuate higher frequencies, the MOV to attenuate impulses, and the capacitor to improve normal mode attenuation. Some are available with single, double, or triple Faraday shielding to improve common-mode attenuation. The major drawback to this type of line conditioner is that it does not regulate voltage.

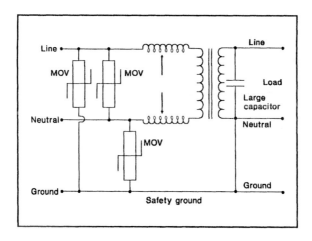

Figure 7-30. Enhanced Isolation Transformer

b. **M-G Sets:** An M-G set, or a motor-generator set, consists of a drive motor mechanically connected to an ac generator. M-G sets generate

301

a new voltage waveform for the load and are therefore somewhat isolated from minor disturbances on the power system.

Motor-generators can shield the load from impulses and from voltage sags and surges. For power line voltage changes of ±20% or more, voltage to the load is maintained at nominal. A useful feature of the motor-generator is its ability to bridge severe short term sags or outages. The momentum of the rotating elements permits the motor-generator to span outages of up to 0.3 seconds (approximately). This time span is mainly limited by the drop in frequency (speed) as energy is removed. The period can be extended by adding inertia via a flywheel as shown in Figure 7-31. Through the combined use of variable-speed, constant frequency or quick-starting engine generator techniques **and** flywheel inertia, it is now practical to extend the time an outage can be spanned so that it is long enough to protect against the common problem of clearing and reclosure caused by feeder faults. Cost, however, is considerably higher with this system than with conventional motor-generators.

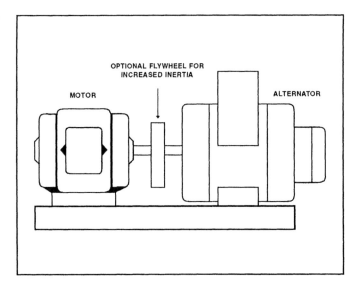

Figure 7-31. Motor-Generator With Flywheel

The problems with motor-generators are mostly on the output or load side. High generator output impedance can cause substantial voltage dips in response to sudden load changes, such as those resulting from large motor starting current. And in addition, response to load changes can be sluggish. Also the drive motor may overheat under long term brownout or low line voltage conditions unless it is oversized. Motor-generator efficiency is typically relatively low, so that electrical energy costs over its lifetime may be substantial. Heat dissipation, weight and bulk, and the potential for audible noise are factors which must be considered in motor-generator installation. Essentially silent machines are available at extra cost. Bearings must be inspected and periodically replaced and/or lubricated in many cases, particularly when flywheels are used. Reliability potential, however, is very high.

Uninterruptible Power Supplies (UPS)

Uninterruptible power supplies are the only solution for continuous operation of computer or other sensitive systems when line voltage interruptions last approximately 0.5 seconds or longer, as is common for a utility. A properly designed UPS system can provide computer-quality power under essentially all normal and abnormal utility power conditions, including outages up to 15 minutes or more

UPS systems are typically solid state, although some are currently made using rotating machinery in combination with solid state conversion. UPS systems have three general configurations illustrated in Figure 7-32.

As can be seen, all the systems contain a battery. The inverter preferred system is essentially the standard configuration for critical equipment because it provides full isolation and power conditioning. The line preferred system (standby) receives its normal flow of power from the utility system and relies on a 1/4 cycle transfer switch to isolate utility system disturbances. This type of system is obviously not as effective as the inverter preferred system and is generally used for small, low cost computer systems where operation is not highly critical.

Line interactive systems, like line preferred systems, have found application in the smaller, less critical sensitive equipment systems. They too change mode of operation upon power failure. They can provide somewhat more power conditioning than the line preferred system, but are inherently the least reliable of the three systems. Normally, power flows through the single-throw static switch and the inductor to the sensitive equipment load and to the converter which, in this mode, acts as a battery charger. By various techniques, the voltage to the sensitive

303

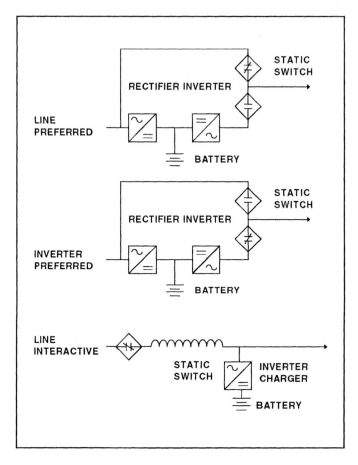

Figure 7-32. UPS Configurations

equipment can be conditioned similarly to the fast response or ferroresonant regulators. Upon failure of the line, the static switch is opened and the function of the converter is changed to that of an inverter delivering power to the sensitive equipment, resulting in a less than a 1/4 cycle disturbance. A <u>line interactive system</u> is vulnerable to loss of output if there is a failure of the converter at any time. Also, taking the converter in and out of service cannot be done without incurring a momentary interruption of voltage to the load. A more recent version of this system provides for isolation of a failed or serviced converter by use

304

of an additional static switch. These problems do not exist with the inverter preferred system.

Again, it should be pointed out that the vast majority of power quality related problems are not caused by the utility but rather are generated on the customers' premises. Some of these, such as spikes or surges, are easily corrected. Others, such as those resulting from power outages, may require costly M-G sets or uninterruptible power supplies (UPS). This so-called conditioned power is not perfect. Even loads connected to the same "conditioned" power supplies interact (see Figure 7-33) and many users have found the need for separate conditioning for each sensitive load.

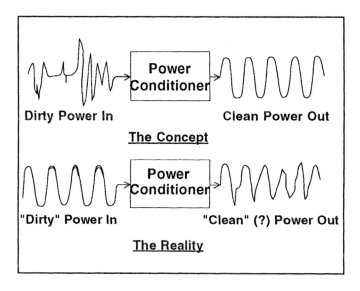

Figure 7-33. Realities of Conditioned Power

ELECTROMAGNETIC FIELDS

Another area sometimes considered a power quality issue is magnetic fields. The question of the health effects of magnetic fields is certainly very complex as well as very confused. It does not appear that conclusive answers will be forthcoming for some time and yet utilities are being asked to investigate ways to mitigate these fields should they be deemed a problem. Although all parts of the utility system create magnetic fields, it is probably the distribution system that causes the most concern because of its proximity, its extensiveness, and the fact that the nature of its design exacerbates the condition.

Measurements of fields, as shown below in Figure 7-34, indicates that magnetic fields resulting from distribution lines are probably lower than other areas of the utility system and considerably lower than those caused by certain appliances. The concern however, is justified mainly on the basis that proximity of people and homes to distribution lines is considerably greater than that of the subtransmission or transmission.

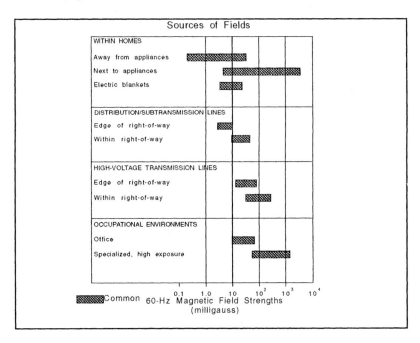

Figure 7-34. Typical Magnetic Field Levels

There are several solutions that have been considered to reduce magnetic fields. While reduction of fields might be a positive outcome, some of these ideas may actually create more severe problems. Some are discussed as follows:

306

Grounded vs. Ungrounded

Distribution systems are usually either delta (ungrounded) or wye (4-wire, multigrounded). The trend is, by far, toward the multigrounded approach. Since a delta system does not utilize the earth as a return path, it has been suggested as one methodology to reduce magnetic fields. While this point cannot be argued, it does present some serious concerns since the present predisposition of utilities toward the grounded system design results from three major factors:

Overcurrent Protection. Overcurrent protection of a distribution system is premised on the fact that short circuit current levels go down as distance from the substation increases. These levels vary from as high as 30K amperes (more likely about 10 kA) to quite low values in the hundreds of amperes. Most of these faults occur as line-to-neutral or line-to-ground faults and fault currents are very high because of the multigrounded neutral design.

A delta system on the other hand produces very little fault current for a line-to-ground fault because there is no zero sequence ground path, and as such produces very low levels of fault current (this due to line capacitance). In any case, it now becomes very difficult for the relay to "see" many of these faults.

Overvoltage Protection. The selection of an arrester rating is based on the system grounding. The system BIL must be, by industry guidelines, at least 20% higher (I believe it should be at least 50% higher) than the protective level of the arrester. The degree of system grounding determines the maximum line-to-ground voltage an arrester will see due to a fault on another phase. At the present time, guidelines suggest the following multiplication factors:

Table 7-3	
System	**Overvoltage Magnitude**
Ungrounded 4-Wire Multigrounded (open wire-gapped)	$1.82 \times E_{LG}$ $1.25 \times E_{LG}$
E_{LG} = Nominal Line-to-Ground Voltage of System.	

An example probably best illustrates the problem. If we had a 34.5, multigrounded system, we would calculate the arrester rating as follows:

$$\frac{34.5 \text{ kV}}{\sqrt{3}} \times 1.25 = 24.9 \text{ kV} => 27 \text{ kV arrester} \qquad \text{7-1}$$

$$\text{IR @ 10,000 amps} \approx 108 \text{ kV}$$

On the other hand, if the system were a delta, we would calculate the arrester rating in this manner:

$$\frac{34.5}{\sqrt{3}} \times 1.82 = 36.2 \text{ kV} => 39 \text{ kV arrester} \qquad \text{7-2}$$

$$\text{IR @ 10,000 amps} \approx 156 \text{ kV}$$

BIL must be, at the very least, 20% greater than the protective level (IR discharge at 10,000 amperes) of the arrester. So we would determine the system BIL as follows:

34.5 kV - Grounded Wye:

$$108 \times 1.20 = 130 \text{ kV} => \text{use 150 kV BIL} \qquad \text{7-3}$$

34.5 kV - Delta:

$$156 \text{ kV} \times 1.20 = 187.2 \text{ kV} => \text{use 200 kV BIL} \qquad \text{7-4}$$

As can be seen, the BIL of the delta system must be higher which of course means higher costs of transformers, lines, cables, switches, etc.

Costs. Besides the cost of higher BIL, there is another major cost that must be considered. Residential load is single phase. To serve this single-phase load requires two wires. With a multigrounded system, these two wires are a phase wire and a neutral wire. On a delta system, two-phase wires must be used. When the residential subdivision is underground, which it usually is these days, the delta system requires two insulated and very expensive cables where the multigrounded system requires only one (and an uninsulated concentric neutral). Other considerations such as more bushing, more disconnects, etc., all increase cost.

Single-Phase vs. Three-Phase

There are utilities in Europe who serve residential loads with three-phase transformers. In the United States, we serve only our large industrial and commercial loads in this manner. It is likely that such a system design reduces magnetic fields because individual phase currents would tend to be lower and the fields would tend to cancel. Current unbalance would also tend to be less. On the other hand, this system would be far too costly to justify in the United States where new residential construction tends to be underground (URD). A comparison of some of the fundamental differences between the two systems is shown below in Table 7-4.

Table 7-4. 3-Phase vs. 1-Phase Residential Distribution	
U.S.	**EUROPE**
• 120/240 • 1-phase transformers heavily overloaded - 25 kVA typical • 4 homes/transformer fairly typical • Higher load density • Fuses are typically expulsion	• 380 Wye/220, 4-wire 416 Wye/240, 4-wire (UK) • Less load per home than U.S. • 3-phase xfrms >>$ 1-phase • Residential units in 300-500 kVA range • 5 to 10 radial, 3-phase, 4-wire secondary feeds, per transformer • No overload • Fuses are current limiting • 100 to 200 dwellings per transformer.

Underground

Underground distribution cables are generally enclosed in plastic duct (CIC) or direct buried. It is of particular interest to note that the earth is *not* a good magnetic field shield at 60 Hz. IEEE Task Force conclusions relative to this subject are:

• Buried pipes often carry sufficient 60 Hz current to substantially change the ground level magnetic field.
• The magnetic fields of buried cables exceed those of an overhead power line carrying the same currents.
• The pipe of a pipe-type cable may be an effective magnetic field shield at 60 Hz.

Line Compaction

Decreasing the spacing between conductors (see Figure 7-35) has been shown to reduce magnetic fields. Another advantage to this compaction on distribution lines is esthetics.

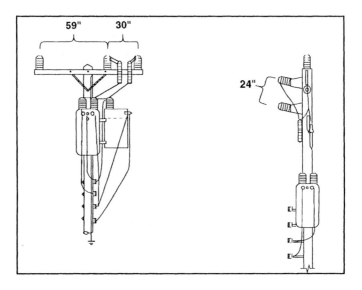

Figure 7-35. Example of Line Compaction

While compaction may present certain advantages, it also presents several of the following disadvantages:

Overvoltage. Bundling or compaction of conductors, such as spacer cable, increases the line-to-ground voltage seen by the unfaulted phases during a line-to-ground fault. This factor changes from 1.25 to 1.5 per unit. This will usually have the effect of causing the multigrounded system to utilize a higher arrester rating, especially for MOV applications. This, in turn, means that the protective margin is decreased should the same BIL be possible to use. Or, in some cases, a higher system BIL may be required.

Overcurrent. Some utilities have found that an instantaneous recloser on a system having reduced phase spacing may not be successful for temporary fault conditions. These utilities have consequently been forced to omit the instantaneous reclosure and utilize a time delayed reclosure after the first instantaneous trip. While this is not necessarily a problem, it does subject the customer to a longer outage (2 to 15 seconds instead of a half cycle) and has been a cause of concern in these days of higher "power quality" and more sensitive loads.

Line Flashovers. Lightning causes many of the temporary faults seen on the distribution system. These faults are a result of the lightning stroke (induced or direct) which causes the insulation (usually air) to break down. A typical overhead line has an insulation level of about 300 kV BIL. Some utilities, in high lightning areas, are seriously considering going to even higher line BILs, such as 450 kV, to decrease the occurrence of line flashover and increase power quality.

Unbalance

Residential loads are single-phase loads and connected phase to neutral on a 4-wire multiground system. In an effort to balance the three-phase loads, single-phase lateral connections are rotated between the three phases. Since it is virtually impossible to completely balance the three phases at all times, unbalances will exist. This three-phase unbalance shows up as residual current and returns to the substation via the neutral wire and the earth. This, of course, adds to the magnetic field since phase cancelization is not as complete.

Typically, this phase unbalance can be significant. Some utilities, for example, set their ground fault relays between 25% and 50% of full load current so as not to misoperate due to system unbalance. This can mean that up to 200 or 300 amperes of unbalanced current can be accommodated before tripping.

Up until recently, it was rather difficult for a utility to monitor unbalance for an extended period of time. Today, however, devices are available which are accurate and convenient to use. Such a device was used to obtain the data shown in Figure 7-36. As can be seen, the feeder monitored here shows very good phase balance for most conditions but some phase unbalance during heavy load periods. It is doubtful whether this condition could be improved to any significant degree by shifting loads amongst phases.

311

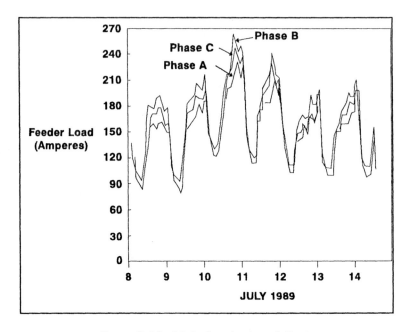

Figure 7-36. Main Feeder Load Cycle

STRAY VOLTAGE

Introduction

Stray voltage is a small voltage of normally less than 10 volts that is usually measured between the earth and neutral. While this voltage could exist on virtually any electrical facility, it is normally associated with dairy farms because farm animals, especially cows, appear to be more sensitive to it then do human beings. Many experts feel that farmers seldom have a stray-voltage problem that affects the health of the animals and/or milk production. They feel that these farmers' problems are generally caused by poor management, improper feeding, poor cleanliness of the animals, improper breeding, defective installation and operation of milking equipment, and poor sanitary conditions. However, very large stray voltages, during experimental tests, have been shown to affect animal behavior. Since utilities can, under certain conditions, contribute to stray voltage, law suits have been common for utilities serving dairy farms.

Sources of Stray Voltage

There are many sources of stray voltage which is one of the reasons it can be so difficult to locate and correct. Several of the more common sources are described as follows:

Bad Connections. Bad connections, such as those caused by corrosion, looseness or even a break, are a primary cause of stray voltage. The reason for this is that this otherwise low impedance path may cause the current to take an unintended path and produce stray voltage.

Improper Wiring. The interconnection of the neutral (white) conductor and the equipment grounding conductor (green or bare wire) is generally required by electrical codes. Compliance with the code implies that the neutral and ground will be at the same potential at these points of interconnection. Improper grounding practices, such as the use of the neutral wire (white) also as the grounding conductor or the use of a ground rod instead of a conductor leading back to the service entrance, may result in stray voltage.

Coupled Circuits. An ac electric field from a high voltage line can capacitively couple to a metallic object, like a metal fence, giving it a measurable voltage.

Utility System Neutral. The neutral of a 3-phase, 4-wire multigrounded (typical distribution system design) utility system will produce a neutral voltage as result of either a line-to-ground fault or unbalanced current on the line. If this system were balanced (same load on all phases), virtually no current would flow in the neutral wire and the earth. However, because this is never the case, some of the current will return via the earth and neutral conductor. A single-phase tap can be considered to be 100% unbalanced since all the return current must return via the earth and neutral conductor.

Figure 7-37 shows how this neutral voltage occurs. As can be seen, the current in this single-phase transformer (or in a single-phase tap) returns to the source via the neutral and earth. Some of the current returns through the earth (we have assumed 40%) and the rest returns through the neutral. The current flowing in the neutral between points A, B, C, and D will produce a voltage on this neutral as a result of Ohm's Law. The voltage at A is then transferred to the secondary neutral via the distribution transformer. This neutral voltage can be different from the customer's earth potential, even though the distribution transformer is grounded, primarily because the ground footing impedance can be high (generally 25 ohms or more). The differential between the neutral voltage and the earth voltage, at the customer's premises, is generally the stray voltage which concerns the utility engineer.

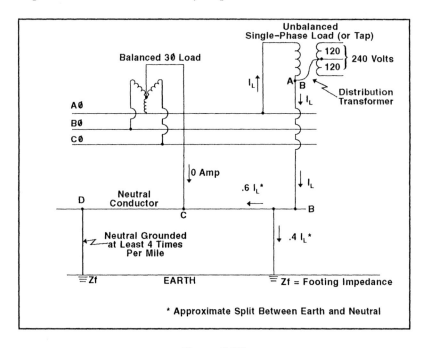

Figure 7-37.

Interconnection of Primary and Secondary Neutrals of Distribution Transformers. The practice of connecting the primary and secondary neutrals on a 4-wire multigrounded system is a matter of safety. If a primary-to-secondary fault developed within the transformer and the neutrals were not connected, the resistance of the return current path could be so high that not enough fault current would flow to enable the primary protective device to clear the fault. High voltage would then be impressed on the secondary for an extended period of time, posing a risk to humans, animals, and equipment. The practice of isolating the secondary neutral is being used on some unigrounded and ungrounded systems. This practice is based upon a concern that a primary fault to the transformer tank could cause a high voltage to be introduced into the secondary utilization services if the primary and/or tank grounds are interconnected to the secondary neutral. This practice is especially not recommended for 4-wire systems.

Mitigation of Stray Voltages. The control and mitigation of stray voltages in livestock facilities demand careful consideration of both animal sensitivity and electrical sources. All mitigation approaches have their own advantages and disadvantages and any approach must consider the alternatives and constraints of a given situation.

Approaches for controlling neutral-to-earth voltages can be categorized as follows:

Voltage Reduction - Reduction of grounding resistance on the distribution neutral will reduce the neutral-to-earth potential due to system loading. Corrective action such as correcting bad neutral connectors, removing faulty loads, balancing loads, improving grounding, or increasing the size of the neutral can be helpful.

Active Voltage Suppression - If modification of the basic electrical system is difficult, one method of mitigation is active voltage suppression. This method utilizes a second source of current to cancel the original source. Disadvantages include the possible maintenance that may be involved, the cost, and the possibility of creating more voltage than needs to be mitigated.

Gradient Control - Gradient control is the same methodology used in substations to control step and touch voltages during faults. Such control by equipotential planes can negate the effects of all neutral-to earth voltages in livestock facilities if it reduces the potential differences at all possible animal contact points to an acceptable level. If an equipotential plane is installed (it should include equipment grounding, metalwork bonding, conductive network in the floor bonded to the electrical grounding), the only possible concern is that animals will receive an electrical shock when they move on or off the plane. One simple and apparently effective solution where it is difficult to retrofit a concrete floor to establish an equipotential plane is the use of insulating paint.

Isolating the Primary and Secondary Neutrals - If the neutral voltage is determined to come from the utility distribution system, one possible solution is a "neutral isolation" device. These devices (see Appendix B, IEEE C62.92.4-1991, *IEEE Guide for the Application of Neutral Grounding in Electrical Utility Systems, Part IV - Distribution*) isolate the primary and secondary neutral for normal load conditions but instantly and solidly connect them together in the event of any system disturbance that causes the voltage of one neutral to rise above a predetermined threshold relative to the other neutral, as could be the case of an internal primary to secondary fault in the transformer.

REFERENCE

1.   L.M. Anderson and K.B. Bowes, "The Effects of Power-Line Disturbances of Consumer Electronic Equipment", '89 TD 423-5 PWRD, Transmission & Distribution Conference, April 1989.

QUESTIONS

1.   Name three conditions that cause power quality problems.

2.   A UPS system will solve most power quality problems. (true or false?)

3.   Name 5 utility power quality problems.

4.   The most important power quality parameter to a utility is:
      a.   primary voltage
      b.   primary current
      c.   secondary voltage
      d.   secondary current

      Explain.

5.   Voltage drop from the feeder to the meter can be as high as _____ volts.

6.	A secondary voltage of 100 volts will cause most VCRs, microwaves and digital clocks to misoperate. (true or false?)

7.	UPS systems can cause more problems than they solve. (true or false?) Explain.

8.	Identify the wave (sag, swell, impulse, etc.), state approximate duration, give a possible cause, and suggest a possible solution for the following:

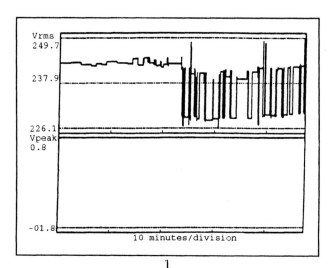

1

8.   (continued)

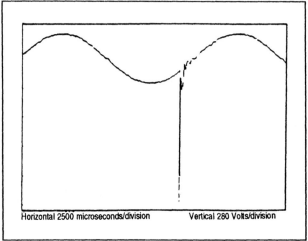

Horizontal 2500 microseconds/division   Vertical 280 Volts/division

2

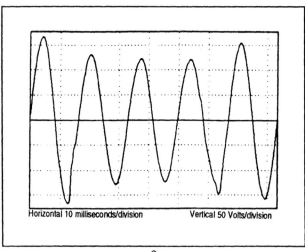

Horizontal 10 milliseconds/division   Vertical 50 Volts/division

3

8.   (continued)

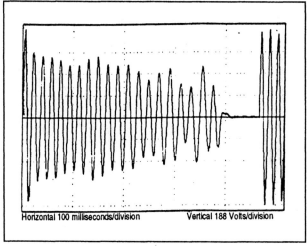

Horizontal 100 milliseconds/division   Vertical 188 Volts/division

4

8

DISTRIBUTION ECONOMICS

INTRODUCTION

It is no longer possible for a utility engineer to justify a design or product change simply on the basis of technical superiority. Today engineers are more often than not placed in the position where they must explain the economic payback of their ideas. In some cases, this is fairly straightforward and conclusive but in others the economics is not that easy to assess.

It is the purpose of this chapter to demonstrate economic techniques that are useful specifically to a utility and are simple to use. These techniques will be developed around such concepts as present worth, carrying charge, cost of losses, and operating costs as well as customer satisfaction.

TIME VALUE OF MONEY

The "time value of money" means that a dollar today has a different value than a dollar a year from now. This is true even without inflation because that dollar given to you today could be invested at a rate higher than inflation. People will pay to use your money. Taking dollars at one point in time and finding its value at some other point in time is called present worth arithmetic. The following is a review of some of the concepts associated with this type of analysis.

Future Worth

The process of taking money and finding its equivalent value at some point in the future is called future worth. For example, one dollar today would be worth 1 dollar and 12 cents a year from now if the established interest rate was 12%. If I wanted to find the value of one dollar two years in the future I would just take the $1.12 I had after the first year and add 12% interest to it. I would now have $1.25 (I would have $1.24 if this were "simple interest"). This concept of future value can be illustrated mathematically as follows.

Here are the principal symbols:

i = Interest rate
n = Number of interest periods
P = Present sum of money
S = Sum of money n interest periods from now.

Then formulas for calculating future worth of money at end of year are:

Year 1 $S = P + iP = P (1 + i)$
Year 2 $S = P (1 + i) + iP(1 + i) = (P + iP) (1 + i) = P (1 + i) (1 + i) = P (1 + i)^2$
Year n $S = P (1 + i)^n$

The amount P is the present value of today's investment while the amount S is the future value. This concept can be shown by a simple diagram, Figure 8-1.

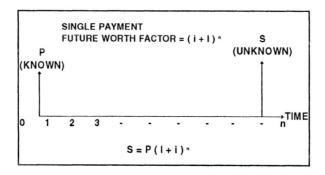

Figure 8-1. Future Worth

An example of the use of the present worth factor follows:

Example 1. How much would you have to invest at 12% interest on January 1, 1985, in order to accumulate $1,791 on January 1, 1991?

Solution 1

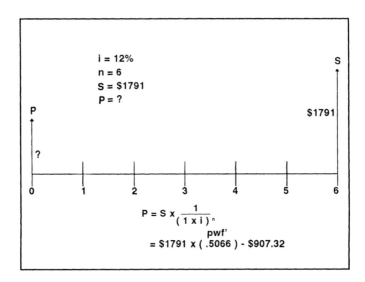

i = 12%
n = 6
S = $1791
P = ?

$$P = S \times \frac{1}{(1 \times i)^n}$$
pwf'
$$= \$1791 \times (.5066) - \$907.32$$

Figure 8-2.

Present Worth

The process of taking future dollars and bringing them back in time, for example back to the present, is called "present worth". Present worth is just the opposite of future worth, so if we take the future worth formula we just developed, we would have the following:

$$S = P(1 + i)^n \qquad \text{8-5}$$

so

$$P = \text{Present Worth} = \frac{S}{((1 + i)^n)} \qquad \text{8-6}$$

An example of the use of the present worth factor (for a single payment follows:

Example 2. What is the present worth on January 1, 1984, of $1,263 on January 1, 1991, if the interest rate is 12% (use Table 8-1)?

Solution 2

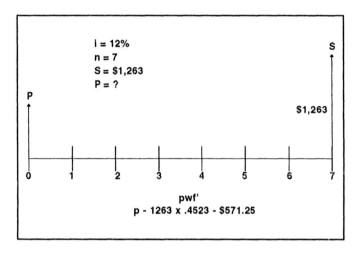

i = 12%
n = 7
S = $1,263
P = ?

pwf'
p - 1263 x .4523 - $571.25

Figure 8-3.

Equal Payments

Utility equipment, not unlike our own possessions, has costs (and savings) associated with them which are calculated on an annual basis. When these costs are uniform, like our mortgage or car payment, we can estimate the value of these payments at some time by using compound interest factors for a "uniform" series of payments. The terms generally used for equal payment economic evaluations are:

1. Sinking fund factor
2. Capital recovery factor
3. Compound amount factor
4. Present worth factor

323

A graphical interpretation of these formulas is shown below in Figure 8-4. These figures are very useful in determining which formula to use. For example, suppose you knew that the application of a capacitor bank could save you $300 per year. If you wanted to know today's value of those yearly savings, you would go to Figure 8-4 and find the graph which fits this evaluation. In this case, we have uniform payments (R) and wish to know the present value (P). We can see that we need to use the <u>present worth factor for a uniform series</u>. The other terms become self explanatory when viewed in relationship to their graphical interpretations.

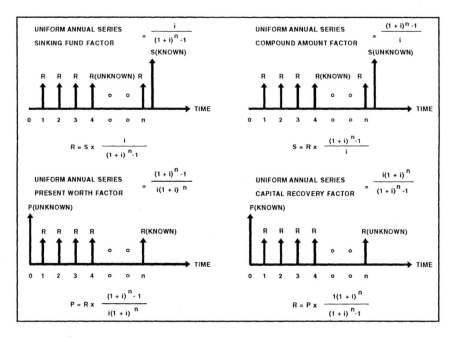

Figure 8-4. Graphical Interpretation of Uniform Series Payments

Example 3. Assume a low loss transformer design can save an estimated $114.10 each year. How much would be saved at the end of 10 years at an interest rate of 12%?

Solution 3

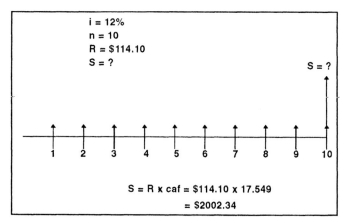

i = 12%
n = 10
R = $114.10
S = ?

S = ?

1 2 3 4 5 6 7 8 9 10

S = R x caf = $114.10 x 17.549
= $2002.34

Figure 8-5.

Example 4. How much would you need to deposit at 12% on January 1, 1991, in order to draw out $179.20 at the end of each year for 7 years, leaving nothing in the fund?

Solution 4

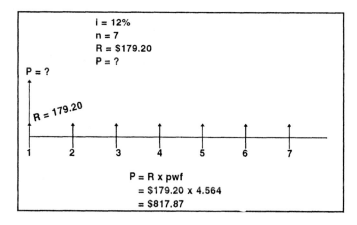

i = 12%
n = 7
R = $179.20
P = ?

P = ?

R = 179.20

1 2 3 4 5 6 7

P = R x pwf
= $179.20 x 4.564
= $817.87

Figure 8-6.

325

TABLE 8-1. 12% COMPOUND INTEREST FACTORS

n	SINGLE PAYMENT		UNIFORM SERIES				n
	Compound Amount Factor caf'	Present Worth Factor pwf'	Sinking Fund Factor sff	Capital Recovery Factor crf	Compound Amount Factor caf	Present Worth Factor pwf	
	Given P To Find S $(1+i)^n$	Given S To find P $\dfrac{1}{(1+i)^n}$	Given S To find R $\dfrac{i}{(1+i)^n-1}$	Given P To find R $\dfrac{i(1+i)^n}{(1+i)^n-1}$	Given R To find S $\dfrac{(1+i)^n-1}{i}$	Given R To find P $\dfrac{(1+i)^n-1}{i(1+i)^n}$	
1	1.120	0.8929	1.00000	1.12000	1.000	0.893	1
2	1.254	0.7972	0.47170	0.59170	2.120	1.690	2
3	1.405	0.7118	0.29635	0.41635	3.374	2.402	3
4	1.574	0.6355	0.20923	0.32923	4.779	3.037	4
5	1.762	0.5674	0.15741	0.27741	6.353	3.605	5
6	1.974	0.5066	0.12323	0.24323	8.115	4.111	6
7	2.211	0.4523	0.09912	0.21912	10.089	4.564	7
8	2.476	0.4039	0.08130	0.20130	12.300	4.968	8
9	2.773	0.3606	0.06768	0.18768	14.776	5.328	9
10	3.106	0.3220	0.05698	0.17698	17.549	5.650	10
11	3.479	0.2875	0.04842	0.16842	20.655	5.938	11
12	3.896	0.2567	0.04144	0.16144	24.133	6.194	12
13	4.363	0.2292	0.03568	0.15568	28.029	6.424	13
14	4.887	0.2046	0.03087	0.15087	32.393	6.628	14
15	5.474	0.1827	0.02682	0.14682	37.280	6.811	15
16	6.130	0.1631	0.02339	0.14339	42.753	6.974	16
17	6.866	0.1456	0.02046	0.14046	48.884	7.120	17
18	7.690	0.1300	0.01794	0.13794	55.750	7.250	18
19	8.613	0.1161	0.01576	0.13576	63.440	7.366	19
20	9.646	0.1037	0.01388	0.13388	72.052	7.469	20

TABLE 8.1. 12% COMPOUND INTEREST FACTORS (CONT)

	SINGLE PAYMENT		UNIFORM SERIES				
n	Compound Amount Factor caf' $\text{Given } P \text{ To Find } S$ $(1+i)^n$	Present Worth Factor pwf' $\text{Given } S \text{ To find } P$ $\dfrac{1}{(1+i)^n}$	Sinking Fund Factor sff $\text{Given } S \text{ To find } R$ $\dfrac{i}{(1+i)^n-1}$	Capital Recovery Factor crf $\text{Given } P \text{ To find } R$ $\dfrac{i(1+i)^n}{(1+i)^n-1}$	Compound Amount Factor caf $\text{Given } R \text{ To find } S$ $\dfrac{(1+i)^n-1}{i}$	Present Worth Factor pwf $\text{Given } R \text{ To find } P$ $\dfrac{(1+i)^n-1}{i(1+i)^n}$	n
21	10.804	0.0926	0.01224	0.13224	81.699	7.562	21
22	12.100	0.0826	0.01081	0.13081	92.502	7.645	22
23	13.552	0.0738	0.00956	0.12956	104.603	7.718	23
24	15.179	0.0659	0.00846	0.12846	118.155	7.784	24
25	17.000	0.0588	0.00750	0.12750	133.334	7.843	25
26	19.040	0.0525	0.00665	0.12665	150.334	7.896	26
27	21.325	0.0469	0.00590	0.12590	169.374	7.943	27
28	23.884	0.0419	0.00524	0.12524	190.699	7.984	28
29	26.750	0.0374	0.00466	0.12466	214.582	8.022	29
30	29.960	0.0334	0.00414	0.12414	241.332	8.055	30
31	33.555	0.0298	0.00369	0.12369	271.292	8.085	31
32	37.582	0.0266	0.00328	0.12328	304.847	8.112	32
33	42.091	0.0238	0.00292	0.12292	342.429	8.135	33
34	47.142	0.0212	0.00260	0.12260	384.520	8.157	34
35	52.799	0.0189	0.00232	0.12232	431.663	8.176	35
40	93.051	0.0107	0.00130	0.12130	767.088	8.244	40
45	163.987	0.0061	0.00074	0.12074	1358.224	8.283	45
50	289.001	0.0035	0.00042	0.12042	2400.008	8.305	50
∞				0.12000		8.333	∞

ANNUAL CARRYING CHARGE

When you buy a car, you not only concern yourself with the cost of the car but also with other things like insurance, interest, gas, and maintenance. Likewise, when a utility purchases equipment, it not only must consider the original cost of the equipment, but also must consider the following:

1. Return on investment
2. Depreciation
3. Income tax
4. Property tax
5. Insurance
6. Operation and maintenance.

The combination of all these costs is called the carrying charge. A brief description of each of these costs is as follows:

Return on Investment

This is the part of the carrying charge which pays investors for the use of their money. It does not consider inflation. A utility has the following two sources of borrowing the money:

• Bonds (or debt money) - repayment of which is called "interest"
• Stock (or equity money) - repayment called "equity return".

The capitalization structure of a utility is the percentage of debt and equity by which the utility borrows its money. For example, if a utility has a capitalization structure of 52% debt (bonds) and 48% equity (stock), its overall return on investment might be calculated as follows:

Example

Given:

Cost of debt money (bonds)   =   9.2%
Cost of equity money (stocks)   =   15.0%

Capitalization structure $= \left\{ \begin{array}{l} 48\% \text{ Equity} \\ 52\% \text{ Debt} \end{array} \right.$

The overall ROI (return of investment) $= .52 \times 9.2 + .48 \times 15.0 = \mathbf{12\%}$

Depreciation

This second major cost component is some times referred to as "recovery of capital" or the "return of investment". Investors not only want interest on their investment but also want that investment back at the end of its life.

There are many different methods that utilities use to depreciate their plant and equipment. The most common is called "straight line" where a fixed percentage of the investment is returned each year until the entire investment is returned. Table 8-2, shown below, is an example of straight-line depreciation for an investment of $10,000 and a life of 5 years.

Table 8-2. Straight-Line Depreciation				
10,000 Investment 5 Year Life Straight Line		Depreciation is returned annually		
Year	Investment	Annual Depreciation	Total Depreciation	ROI
1	10,000	2,000	2,000	1,200
2	8,000	2,000	4,000	960
3	6,000	2,000	6,000	720
4	4,000	2,000	8,000	480
5	2,000	2,000	10,000	240

Federal Income Tax

Equity return (stock dividends) are a cost of borrowed money because they are not deductible in computing taxes. When a utility invests in a plant it, in a sense, incurs a tax obligation for future years on that plant. So, when a utility borrows money from investors, it must pay not only the investors but also the federal government almost as much, which can be illustrated as follows:

- P = Profit = Total revenue required to pay investors and taxes
- Assume tax rate is 48% of total profit.
- Assume investors (equity) earn $1, then

$$P = \text{Equity Return} + \text{Taxes}$$

$$P = 1 + .48 \ (P) \qquad\qquad 8\text{-}3$$

$$=> P = 1.92$$

OR

That for every dollar we pay investors we need $1.92 in profit

OR

For every $1 of equity return I need $.92 for the feds.

Property Taxes

Annual property taxes may be expressed as a percentage of the capital investment. This is of the order of 2.5%.

Insurance

Plant insurance is also an expense that may be directly related to the capital investment in a plant and may be expressed as a percentage thereof. Insurance might be 0.25% per year.

Operating and Maintenance Expenses

The calculation of this cost component varies with the nature of the project. It is usually not a direct function of the capital invested and may have an inverse tendency. That is, alternatives often exist for higher capital expenditures to reduce operating costs. Therefore, it is not expressed as a percent of capital investment in most cases.

Calculation of Annual Carrying Charges

The annual carrying charge is the summation of all these costs. It is most convenient to have a single percentage number to represent the annual cost of return on investment, depreciation, federal income tax, property tax, and insurance. Such a number is called the levelized annual carrying charge. Multiplying the capital investment by the levelized annual carrying charge will give the annual cost of the investment in the system.

To illustrate the calculation of carrying charges, the following table is proposed:

Table 8-3. $100 Investment - 5 Year Life, Straight Line Depreciation, 48% Federal Income Tax					
	YEAR				
	1st	2nd	3rd	4th	5th
Investment for the year	100	80	60	40	20
Return Equity (7.2%)	7.20	5.76	4.32	2.88	1.44
Interest (4.8%)	4.80	3.84	2.88	1.92	.96
Depreciation	20.00	20.00	20.00	20.00	20.00
Income Tax	6.65	5.32	3.99	2.66	1.33
Yearly Charge	**38.65**	**34.92**	**31.19**	**27.46**	**23.73**

To convert these yearly charges to a single annual carrying charge, the present worth of the five yearly charges must be calculated (using a 12% rate of return) and then converted to an equivalent uniform annual carrying charge. This is shown as follows:

$$
\begin{aligned}
&= (38.65 \times .8929 + 34.92 \times .7972 + 31.19 \\
&\quad \times .7118 + 27.46 \times .6355 + 23.73 \times .5674) \times .27741 \\
&= (34.51 + 27.84 + 22.20 + 17.54 + 13.46) \times 0.27741 \\
&= 32.03
\end{aligned}
\qquad 8\text{-}4
$$

To this must be added the annual cost of property tax and insurance. The total annual carrying charge for the capital investment is:

Return on Investment ⎫
Depreciation   ⎬ 32.03
Federal Income Tax ⎭
Property Tax   2.50
Insurance   .25
                      34.78

Annual carrying charges get smaller as equipment life gets longer. The total carrying charges for various service life and 12% return are:

Years of Life	Total Levelized Annual Carrying Charge
5	34.78%
10	24.71%
15	21.87%
20	20.78%
25	20.35%
30	20.19%

DEFINITION OF TERMS

Diversity

Diversity is a term used to cover the fact that individual loads occur at different times. This means that if the maximum load of two or more loads are added, their sum will generally be greater than the true sum because these peaks occur at different times. Figure 8-7, below, shows a typical diversity for three types of residential loads. These loads have been labeled "high use" or all electric, "medium use" or full use where the customer has all the major appliances, and "low use" like you might see for an apartment dweller. Use of these curves is fairly straightforward. For example, suppose a transformer is needed to supply 6 all electric or "high use" homes. The maximum demand per home is about 20 kVA.

The diversified demand per home for 6 homes is about 13 kVA, so we would need a transformer that could handle 78 kVA at peak.

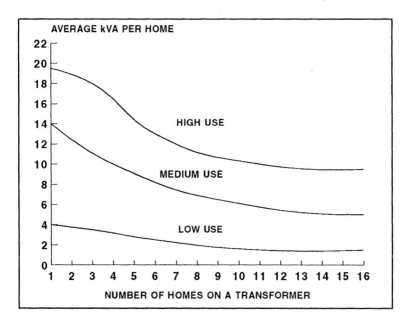

Figure 8-7. Typical Diversity Curve

Load Factor

Load factor is defined as the ratio of the average load over a given period of time to the peak load during that time. This factor can be determined very simply by dividing the number of kw-hr metered during the period by the product of the number of hours in the period times the peak kilowatts. This is an important characteristic for the economic well-being of the electric utility company.

In effect, the peak load determines how much system capacity and, hence, system investment is required to serve a particular load or group of loads, where as the average load determines the kw-hr billing revenue which will be obtained from serving that same load. Hence, high load factor is favorable and low load factor unfavorable to the utility's economic welfare.

333

Overall system load factor of individual companies vary significantly from the national average of about 60%. Figure 8-8 shows actual load factor curves for two utility companies. Company A enjoys annual load factors higher than the national average, while Company B has load factors markedly below the national average.

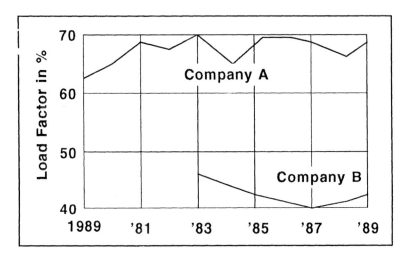

Figure 8-8. Individual Company Load Factors

Loss Factor

Loss factor is defined as the ratio of the average losses over a specified period of time to the peak losses during that same period. While it is relatively easy to determine load factor, it is difficult to determine loss factor because average system losses cannot be easily determined. The reason for this is, in part, due to the fact that system losses vary as the square of the current. The evaluation of energy losses is of prime importance, for it represents the loss of a real, saleable product.

As would be expected, "loss factor" varies as some function of "load factor". The following empirical relationship between load factor (Ld F) and loss factor (Ls F) has been determined to be a relatively good approximation for calculations on a distribution system having a typical load cycle:

$$Ls\ F\ =\ 0.15\ Ld\ F\ +\ 0.85\ Ld\ F^2 \qquad 8\text{-}5$$

The cost of conductor losses per year = (loss factor) (kW loss at peak load) (cost per kW-hr) (8760).

Transformer core losses, while relatively small in comparison with the total conductor losses, should not be neglected, and should, of course, be evaluated on the basis of 100 percent loss factor or equivalent hours.

Responsibility Factors

Responsibility factor is the ratio of that part of an individual load that occurs at the time of the system peak load to the individual load peak. It is thus a measure of how much the individual load contributes to the system peak load. Typical peak responsibility factors which can be used for evaluating the cost of losses in various parts of the system are shown in Table 8-4.

Table 8-4			
LOSSES			
	Primary	**Transformer**	**Secondary**
To Production Peak	0.58	0.56	0.56
To Transmission & Station	0.60	0.58	0.58
To Primary Peak	1.00	0.85	0.85
To Transformer Peak	--	1.00	1.00
To Secondary Peak	--	--	1.00

ECONOMIC EVALUATIONS

Economic evaluations are made in as almost endless variety of ways. The approaches for making economic evaluations used in this text follow the philosophy below:

a. Estimate the duration of the comparison. If two devices are being compared and they have the same life, then only one life cycle is normally used. If they have different service lives, then it is necessary to go through enough life cycles until the alternatives terminate at the same time. If this doesn't happen, then taking the evaluation out to 60 years or more is sufficient.

b. Calculate the annual costs associated with each device. For example, if we purchased a piece of equipment, we would have to pay a levelized annual carrying charge on this equipment.

c. Calculate their present sum and sum up for each scheme.

Examples are shown below to illustrate different approaches for making economic comparisons:

Switches

Suppose you had a choice between two manufacturer's switches. The first manufacturer offered an oil switch for a cost of $12,000 and you knew it had an annual maintenance cost of $50 per year and an average life of 20 years. The second manufacturer offered an SF_6 switch for $15,000. He claimed that this switch had an expected life of at least 30 years and an average annual maintenance cost of $25. The cost of money is 12%. You are asked to determine which switch is the least expensive.

The total time period for the evaluation is 60 years because this number is divisible by both the 20 year life and the 30 year life. The evaluation is as follows:

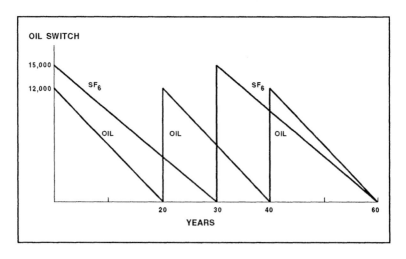

Figure 8-9. Life Cycle for Economic Comparison

Solution

1.   Calculate Cost of Oil Switch

The carrying charge for a 20 year life has been previously calculated to be 20.78%. The present worth of the 3 switches over the 60 year period is calculated as follows:

0-20 (Cost)(Carrying Charge)(pwf) ($12,000)(.2078)(7.469) = $18,624.70
20-40 (Bring to present from year 20) $18,624.70 * .1037 =   1,931.38
40-60 (Bring to present from year 40) $18,624.70 * .0107 =   199.28
Switch Total $20,755.36
P.W. cost of maintenance over 60 years $50 x 8.333 =   416.65
Total Cost of Oil Switch $21,172.01

2.   Calculation of SF$_6$ Switch

0-30 ($15,000)(.2019)(8.055) = $ 24,394.57
30-60 ($24,394)(.0334) =   814.76
0+M $25 x 8.333 =   208.33
Total Cost of SF$_6$ Switch $ 25,417.66

∴ The oil switch would be more cost effective.

Capacitors

A 10 mile, 13.8 kV utility line with an impedance of .2 + j.6 ohm/mile serves an industrial customer with a peak load of 8 MVA at .8 PF and a .6 load factor. The customer has just purchased a 3 MVAR bank of capacitors. Is the cost of losses reduced at a rate of $.08 per kHR?

Solution

1. Losses without caps

$$\frac{8 \text{ MVA}}{3 \times \frac{13.8}{\sqrt{3}}} = 334 \text{ Amps Peak} \qquad \text{8-6}$$

$$\text{Loss Factor} = \frac{\text{Average Loss}}{\text{Peak Loss}}$$

$$= .15 \times \text{Load Factor} + .85 \times (\text{Load Factor})^2$$

$$= .15(.6) + .85(.6)^2 = .396 \qquad \text{8-7}$$

$$\text{Avg. Loss} = I^2 R \times \text{Loss Factor}$$

$$= (334)^2 \times (10 \times .2) \times .396$$

$$= 88.35 \text{ kW Per Phase}$$

$$\text{Total Annual Losses} =$$
$$3 \times 8760 \times 88.35 = 2,321,838 \text{ kWHR} \qquad \text{8-8}$$
$$\text{Cost/Year} = \$185,747/\text{Year}$$

2. Losses with capacitor bank

$$8 \text{ MVA at } .8 \text{ PF} = 6.4 \text{ MW} \qquad \text{8-9}$$
$$= 4.8 \text{ MVAR}$$

With capacitor load

$$= 6.4 \text{ MW and } 1.8 \text{ MVAR} \qquad \text{8-10}$$
$$= \text{New Load is } 6.65 \text{ MVA}$$

$$I = \frac{6.65}{\sqrt{3} \times 13.8} = 278.22 \text{ Amps} \qquad \text{8-11}$$

$$\text{Avg. Loss} = (278.22)^2 (10 \times .2) \times .396 \qquad \text{8-12}$$
$$= 61.31 \text{ kW Per Phase}$$

Total annual cost of losses =

$$3 \times 8760 \times 61.31 \times .08 = \$128,898 \qquad \text{8-13}$$

Cost Savings:

$$\begin{array}{r} \$185,747 \\ \underline{-128,898} \\ \$\ 56,849 \end{array}$$

Transformers

Economic evaluation of transformers involves several of the concepts discussed earlier. To properly evaluate the cost of owning a transformer, the following three annual costs must be considered:

1. Annual carrying charge of the unit
2. Cost of no-load losses
3. Cost of load losses.

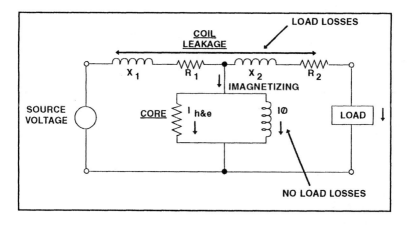

Figure 8-10. Transformer Model

The losses in a transformer occur under both loaded and unloaded conditions. These losses can be further broken down into two major parts:

1. Energy component or production cost to generate kWhr losses.
2. Demand component or annual costs associated with system investment required to supply the peak kW of loss.

As we can see in the figure above, the core of the transformer always has approximately the same voltage across it and consequently the I^2R losses in the core (magnetizing branch) are constant, occurring 24 hours a day, 365 days a year. Since these losses occur even when there is no load on the transformer, they are called "no load losses". Also, since these losses occur at system peak, the utility must install a kW of system capability for every kW of no load loss (responsibility factor is 1).

The "load losses" are the losses in the leakage impedance related to load conditions. These losses, sometimes referred to as copper losses even though most windings are made from aluminum, go up as the square of the load. Because the peak load on a transformer may occur at a different time than the system peak, it is not usually necessary to build a kW of system for every kW loss due to load. Consequently, the responsibility factor is usually less than 1.

Example. There is a 5.6 kVA single-phase secondary load with a growth rate of 6% to be served. A 15 kVA distribution transformer is readily available. How much extra in transformer cost and loss evaluation would it be to use this transformer rather than a 10 kVA? Both transformers have a 20 year life, and operating and maintenance expenses are assumed to be the same. Assume the following values:

System cost   = \$800/kW
Energy   = \$0.03/kWh
Responsibility factor   = 0.9
Loss factor   = 0.15
Transformer costs 10 kVA = \$150
Transformer costs 15 kVA = \$200

Losses =	TRANS.	LOAD	NO LOAD
	10 kVA	125 W	59 W
	15 kVA	179 W	76 W

Formulas for determining transformer loss evaluations are:

Annual Cost No Load Power Loss

$$= (P + 8760E) \text{ No Load Loss} \qquad \text{8-14}$$

Annual Cost Load Power Loss

$$= K^2 (P \times RF + 8760 \times L_sF \times E) \text{ Load Loss} \qquad \text{8-15}$$

K   = Equivalent annual peak load
RF   = Responsibility factor
L_sF = Loss factor
P   = Annual cost of system capacity
E   = Energy cost ($/kWh)

The first thing to do is to determine K, which can be done as follows:

341

Table 8-6. 10 kVA Transformer Loading				
Year	PU	PU^2	12% PVF	PV
0	0.560	0.314	1.0000	0.314
1	0.594	0.353	0.8929	0.315
2	0.629	0.396	0.7972	0.316
3	0.667	0.445	0.7118	0.317
4	0.707	0.500	0.6355	0.318
5	0.749	0.561	0.5674	0.318
6	0.795	0.632	0.5066	0.320
7	0.842	0.709	0.4523	0.321
8	0.893	0.797	0.4039	0.322
9	0.946	0.895	0.3606	0.323
10	1.003	1.006	0.3220	0.324
11	1.063	1.130	0.2875	0.325
12	1.127	1.270	0.2567	0.326
13	1.194	1.426	0.2292	0.327
14	1.266	1.603	0.2046	0.328
15	1.342	1.801	0.1827	0.329
16	1.422	2.022	0.1631	0.330
17	1.508	2.274	0.1456	0.331
18	1.598	2.554	0.1300	0.332
19	1.695	2.873	0.1161	0.334
				6.47

Uniform Annual Series = $0.13388 \times 6.806 = 0.8662$

Therefore, for the 10 kVA transformer $K^2 = 0.8662$

Since per unit loading of the 15 kVA transformer is $\left(\dfrac{10}{15}\right)$ of the 10 kVA transformer, K^2 for the 15 kVA transformer is 0.3850 (i.e., $(2/3)^2 \times 0.8662$).

10 kVA Transformer

Annual Cost of Transformer 150 x .2078	=	31.17
Load Loss = 0.8662 (0.9 x 800 x .2078 + 8760 x 0.15 x 0.03)0.125	=	20.46
No Load Loss = (800 x .2078 + 8760 x 0.03)0.059	=	25.31
Total	=	76.94

15 kVA Transformer

Annual Cost of Transformer 200 x .2078	=	41.56
Load Loss = 0.3850 (0.9 x 800 x .2078 + 8760 x 0.15 x 0.03)0.179	=	13.02
No Load Loss = (800 x .2078 + 8760 x 0.03)0.076	=	32.61
Total	=	87.25

It would cost $9.66 more per year to use the 15 kVA transformer rather than the 10 kVA.

Automation

The methodology utilized to evaluate the value of a function is normally performed with consideration of both the carrying charge associated with the additional necessary equipment and the rate of return associated with the appropriate savings in losses, etc. The following is an example of a methodology used by the author to evaluate these relative values to indicate whether a function is justified.

Assume that there are two feeders which are equally loaded for approximately 21 hours of each day but feeder A has a 3 hour peak which occurs when feeder B is "off peak". The normal configuration of this feeder for this peak period on feeder A is shown in Figure 8-11. Also assume that the 5 mile section of line at the end of feeder A contains 200 amps of load.

As can be seen, the feeders for this hypothetical (and greatly exaggerated) case are unevenly loaded during this period. An additional breaker could be installed 10 miles out on feeder A to balance these loads as shown in Figure 8-12.

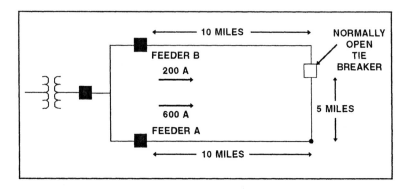

Figure 8-11. Normal Configuration at Feeder A Peak

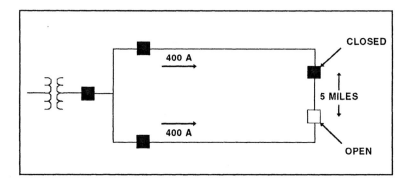

Figure 8-12. Automated Configuration at Feeder A Peak

If the loads are considered to be evenly distributed and the wire resistance is the same in all sections of the system, the percent reduction in losses can be calculated as follows:

Old System:

$$\text{Losses} = L = I^2R = (\frac{I^2R}{3} \text{ for evenly distributed load}) \qquad 8\text{-}17$$

$$L = 3 \times \frac{1}{3} [(200)^2 \times R + (600)^2 \times R + (200)^2 \times \frac{R}{2}]$$

$$= (200)^2 + (600)^2 + \frac{(200)^2}{2} \times R \qquad \text{8-18}$$

$$= 420{,}000 \times R$$

New System:

$$L = (400)^2 \times R + (400)^2 \times R + (200)^2 \times \frac{R}{2} \qquad \text{8-19}$$

$$= 340{,}000 \times R$$

% Change in Losses =

$$\frac{420{,}000 - 340{,}000}{420{,}000} \times 100\% = 19\% \qquad \text{8-20}$$

Let R be equal to 5 ohms (.5 ohms/mile × 10 miles). The change in losses is then equal to:

$$\frac{\dfrac{(420{,}000 - 340{,}000) \times 5}{1000 \text{ Watts}}}{kW} = 400 \text{ kW} \qquad \text{8-21}$$

Assume that the load factor for the peak 3 hour period is 90%. Then the average losses can be calculated as follows:

$$\text{Loss Factor} = \frac{\text{Average Loss}}{\text{Peak Loss}}$$

$$\text{Loss Factor} = .15 \text{ (Load Factor)} + .85 \text{ (Load Factor)}^2$$
$$= .15 \ (.9) + .85 \ (.9)^2 = .135 + .6885$$
$$= .82$$

$$\text{Average Loss} = \text{Loss Factor} \cdot \text{Peak Loss}$$
$$= .82 \times 400 = 328 \text{ kW}$$

$$\text{Kilowatt Hours} = 3 \text{ hrs/day} \times \frac{261 \text{ Weekdays}}{\text{Year}} \times \text{kW}$$
$$= 3 \times 261 \text{ Weekdays} \times 328$$
$$= 256,824 \text{ kWhr}$$

8-22

The cost savings per year at 5 cents per kilowatt hour is equal to $12,841.

Suppose the cost to implement this additional function of load switching is $50,000 for the additional breaker and modifications to the existing open tie point. The question is, how do we estimate whether this additional automated function makes economic sense?

Straight present worth analysis is not completely accurate since the total revenue requirements of the new equipment must include such considerations as:

• Return on investment
• Depreciation
• Income tax
• Property tax
• Insurance
• Operating and maintenance expenses.

These costs change from year to year but are most commonly represented by a levelized value normally referred to as an annual carrying charge. If we assume that the additional equipment has a life of 20 years, a levelized annual carrying charge of 20%, and that the cost of money (return) is 12%, then the total present worth cost of ownership can be calculated as follows:

$$\text{Annual Carrying Charge} = \$50,000 \times .20 \qquad \text{8-23}$$
$$= \$10,000$$

P = Present worth
R = Uniform annual cost
η = Years

$$P = R \times \frac{(1 + \iota)^{\eta} - 1}{\iota (1 + \iota)^{\eta}} \qquad \text{8-24}$$

Where ι = 12% and η = 20 Years
P = 10,000 × 7,469
P = 74,690 = Cost of System Changes

The present worth value of the losses saved each year can be calculated in a similar manner as follows:

$$P = \$12,841/\text{yr.} \times 7.469 \qquad \text{8-25}$$
$$= \$95,909$$

Consequently, the addition of this function will save the utility approximately $21,000 over the life of the equipment.

REFERENCE

1. Campbell, H.E., "Distribution Economics", Chapter 3, Power Distribution System Course, PTI-1982.

QUESTIONS

1. If we have no inflation, then $500 today is equal to $500 in 5 years. (true or false?)

2. If 12% is the established annual simple interest rate, then $100 today is equivalent to $____ 3 years from now.

3. If 12% is the established annual compounded interest rate, the $100 today is equivalent to $_____ 3 years from now.

4. Why must a utility maintain a good ROI (return on investment)?

5. "Return" is the total depreciation value of the plant. (true or false?)

6. The cost of stock money is called interest. (true or false?)

7. The return of capital or return of investment is referred to as _____.

8. Carrying charge is composed of what areas?

9. Load factor is defined as the ratio of the average power loss over a designated period of time to the maximum loss occurring in that period. (true or false?)

10. The cost of transformer losses can be broken down into what two categories?

11. The no load losses of a 10 kVA transformer are _____ (higher, lower) than a 37.5 kVA transformer?

12. The total losses of a 10 kVA transformer are always less than a 37.5 kVA transformer. (true or false?)

13. What three factors should be considered in the evaluation of the total investment in a distribution transformer?

14. When economically comparing two different schemes having different service lives, what must you do?

INDEX